Imperfect Information and Investor Heterogeneity in the Bond Market

Contributions to Economics

Christoph M. Schneider
Research and Development Management: From the Soviet Union to Russia
1994. ISBN 3-7908-0757-5

Lars Olof Persson/Ulf Wiberg
Microregional Fragmentation
1995. ISBN 3-7908-0855-5

Ernesto Felli/Furio C. Rosati/ Giovanni Tria (Eds.)
The Service Sector: Productivity and Growth
1995. ISBN 3-7908-0875-X

Giovanni Galizzi/ Luciano Venturini (Eds.)
Economics of Innovation: The Case of Food Industry
1996. ISBN 3-7908-0911-X

David T. Johnson
Poverty, Inequality and Social Welfare in Australia
1996. ISBN 3-7908-0942-X

Rongxing Guo
Border-Regional Economics
1996. ISBN 3-7908-0943-8

Oliver Fratzscher
The Political Economy of Trade Integration
1996. ISBN 3-7908-0945-4

Ulrich Landwehr
Industrial Mobility and Public Policy
1996. ISBN 3-7908-0949-7

Arnold Picot/Ekkehard Schlicht (Eds.)
Firms, Markets, and Contracts
1996. Corr. 2nd printing 1997
ISBN 3-7908-0947-0

Thorsten Wichmann
Agricultural Technical Progress and the Development of a Dual Economy
1997. ISBN 3-7908-0960-8

Ulrich Woitek
Business Cycles
1997. ISBN 3-7908-0997-7

Michael Carlberg
International Economic Growth
1997. ISBN 3-7908-0995-0

Massimo Filippini
Elements of the Swiss Market for Electricity
1997. ISBN 3-7908-0996-9

Frank Hoster/Heinz Welsch/ Christoph Böhringer
CO_2 Abatement and Economic Structural Change in the European Internal Market
1997. ISBN 3-7908-1020-7

Christian M. Hafner
Nonlinear Time Series Analysis with Applications to Foreign Exchange Rate Volatility
1997. ISBN 3-7908-1041-X

Sardar M.N. Islam
Mathematical Economics of Multi-Level Optimisation
1998. ISBN 3-7908-1050-9

Sven-Morten Mentzel
Real Exchange Rate Movements
1998. ISBN 3-7908-1081-9

Lei Delsen/Eelke de Jong (Eds.)
The German and Dutch Economies
1998. ISBN 3-7908-1064-9

Mark Weder
Business Cycle Models with Indeterminacy
1998. ISBN 3-7908-1078-9

Tor Rødseth (Ed.)
Models for Multispecies Management
1998. ISBN 3-7908-1001-0

Michael Carlberg
Intertemporal Macroeconomics
1998. ISBN 3-7908-1096-7

continued on page 117

Frank Riedel

Imperfect Information and Investor Heterogeneity in the Bond Market

With 7 Figures

Physica-Verlag
A Springer-Verlag Company

Series Editors
Werner A. Müller
Marina Bihn

Author
Dr. Frank Riedel
Department of Economics
Institute of Economic Theory
Humboldt University
Spandauer Straße 1
D-10178 Berlin
Germany

ISBN 3-7908-1247-1 Physica-Verlag Heidelberg New York

Cataloging-in-Publication Data applied for
Die Deutsche Bibliothek – CIP-Einheitsaufnahme
Riedel, Frank: Imperfect information and investor heterogeneity in the bond market / Frank Riedel. – Heidelberg: Physica-Verl., 2000
(Contributions to economics)
ISBN 3-7908-1247-1

Softcover Design: Erich Kirchner, Heidelberg

SPIN 10744575 88/2202-5 4 3 2 1 0 – Printed on acid-free paper

Acknowledgements

I am particularly grateful to Prof. Dr. Hans Föllmer, my supervisor, who gave me the opportunity to experience outstanding and inspiring research. Discussions with him were an encouraging and motivating help when I started this work.

Being a member of the Graduate School of Applied Microeconomics at Humboldt University was a perfect basis for preparing this book. The joint lecture about current problems of microeconomic theory by Prof. Dr. Bester, Prof. Dr. Güth and Prof. Dr. Wolfstetter contributed essentially to my conversion from a mathematician into an economist. I am especially indebted to Prof. Dr. Elmar Wolfstetter for his continuous support and advice.

Many participants of the Graduate School gave useful comments. It is a pleasure to mention Dr. Ida Bagel, Dr. Uwe Dulleck, Dr. Henning Dankenbring, Prof.Dr. Ulrich Kamecke, Michael Kaul, Prof.Dr. Alan Kirman, Prof. Dr. Lutz Kruschwitz, Dr.Dr. Andras Loeffler, and Jörg Oechssler, Ph. D.

I appreciate the grammatical and linguistic help by Milto Kitsios and Katrin Riedel.

Financial support by Deutsche Forschungsgemeinschaft, Graduiertenkolleg Angewandte Mikroökonomik is gratefully acknowledged.

Last not least I thank my parents Anne and Norbert Riedel for their generous support over the years. This book is dedicated to them.

Berlin, May 1999 *Frank Riedel*

Contents

Introduction

On today's asset markets, investors can choose among a wide range of debt instruments with distinct effective yields. In particular, interest rates differ for different *terms*, that is, times until maturity. The relationship between term and interest rate is called the term structure of interest rates. Usually, the longer the term, the higher is the effective yield, and the term structure is rising. However, sometimes the converse occurs, and yields for long term instruments are lower than those for short term instruments, implying declining or humped term structure shapes. It is therefore important to understand why the term structure can have these different forms and how various factors influence the term structure of interest rates.

In the present book, two possible causes for distinct term structures are studied: Imperfect information and investor heterogeneity. Imperfect information arises, for example, from the fact that investors do not know the growth rate or other significant parameters of the economy. Instead, these parameters must be estimated on the basis of the observable data and investment decisions are based upon the estimates. Heterogeneity plays an important role on the asset market, as every practitioner knows. In theoretical models, however, homogeneous markets are frequently studied in which all agents share the same characteristics.

Literature

The theoretical literature[1] can be divided into two periods, the periods before and after the scientific revolution initiated by the work of R. Merton [47] and F.Black and M.Scholes[8].

In the writings of Lutz and more explicitly in those of Malkiel [42] written in the traditional period before said revolution, one finds the relationships that connect the different interest rates in a world of certainty for no-arbitrage reasons. Lutz and Malkiel also take the influence of uncertainty into account although they do not have the necessary tools required to treat such a case formally. Instead of presenting a formal model, several heuristic arguments are used to give support to the various forms of the Expectations Hypothesis.

The Expectations Hypothesis asserts that, to paraphrase, the relations which hold true in a world of certainty should also hold true in an uncertain world, at least in expectation. This idea can be expressed in various ways. One version of the Expectations Hypothesis states that expected yields of different bonds should be equal, another one, that the forward rate is equal to the expected future spot rate[2].

The idea that expected yields of different bonds should be equal corresponds, at least heuristically, to risk neutrality. To include risk aversion, the Expectations Hypothesis is extended to include some kind of premia for holding more risky assets. This is then called the Liquidity Premium Hypothesis. It is inspired by Keynes [35][3] and his Liquidity Preference Theory which asserts that, the less liquid a commodity or an asset is, the higher must be its own rate of interest. Applying this idea to the bond market, Hicks [27] claims that most investors prefer to lend short and to borrow long, because they prefer "to hold their money on deposit in some form or other"

[1] I restrict myself to the theoretical work on the term structure. A survey of the empirical literature is contained in the Handbook article written by Shiller[56].

[2] Although the various forms of the Expectations Hypothesis are not distinguished in the classical writings, they are not consistent with each other as Cox, Ingersoll and Ross [10] show.

[3] see Chapter 17, part II, of the *General Theory*.

(p.190). Thus, there would be a higher demand for short term bonds than for long term bonds leading to higher long term than short term yields. Lutz [41] combines this idea of liquidity preference with the one of risk aversion to obtain the same conclusion. The longer the term of an instrument, he says, the greater is the risk to loose money due to a change in interest rates. Therefore, investors may require a positive risk or term premium for buying long term bonds.

Modigliani and Sutch [50] criticized the Hicks-Lutz prediction of positive term premia by pointing out that it is not necessarily rational for an investor to prefer short to long term instruments. For example, if an investor is saving for a child's education ten or twenty years ahead, then buying a bond with ten or twenty years to maturity is less risky than rolling over the money at random short rates. Therefore, the 'preferred habitat' of an agent plays an important part in determining the demand for bonds. If many people share a long habitat, negative term premia are possible, since the demand for long term bonds is then higher than the demand for short term instruments. Thus, time preferences of the investors (their habitats) determine the sign of the term premium, which in contrast to the Liquidity Premium Hypothesis can be positive or negative, depending on the distribution of time preferences in the economy. The 'Preferred Habitat Hypothesis', as it is called, inspired Chapter 2 herein.

After Merton [47] and Black and Scholes [8] drastically changed the theory on financial markets, the seventies saw the first no-arbitrage and partial equilibrium models of the term structure, where the continuous-time nature of the asset market is formally modeled. Endowed with the no-arbitrage arguments developed by Black and Scholes and the formalism of stochastic calculus introduced by Merton, researchers could start studying the consequences of the concept of no-arbitrage for the term structure. Merton himself, in [46] and [48],Chapter 11, as well as Vasiček[61], Dothan [17] and Richard [54] obtained closed-form solutions for the term structure under various assumptions about the dynamics of the short rate.

In the eighties, a whole range of literature emerged looking for "good" no-arbitrage models of the term structure (for a survey, Rogers [55]). In these models, the price dynamics for a set of suitably chosen bonds or likewise, interest rates, are the starting point rather than the result. The aim is to develop a good model in the sense that derivative pricing is possible and a good fit to empirical prices is simultaneously obtained.

The focus in the present book, though, is on equilibrium models, in which bond prices are determined endogenously. This approach is partially pursued in the early literature by Merton [48] and Vasiček [61] and has been fully developed in the famous contribution of Cox, Ingersoll and Ross [11]. They examine a factor-model of a production economy with constant returns to scale production opportunities. Another equilibrium approach, followed by Bakshi, Chen [7], Duffie, Schroder, Skiadas [22] and Sun [58], is to consider pure exchange economies in the spirit of Lucas[40]. In these models, agents' consumption endowment is the primitive which is exogenously given.

Most general equilibrium models of the term structure of interest rates have the disadvantage that they make specific parametric assumptions about the preferences of the representative investor. In principle, all the models cited above describe a one-investor economy or an economy in which all investors are identical in their preferences, beliefs and endowments.

An exception to this are the articles of Dumas [23] and Wang [62] as well as that of Detemple and Murthy [15]. Dumas and Wang examine the impact of differing degrees of risk aversion among investors on the term structure. Detemple and Murthy allow for heterogeneous beliefs. In two chapters of the present book, Chapters 2 and 4, the strong homogeneity assumption is also relaxed and the effect of investor heterogeneity on the term structure is studied as I will explain below.

Plan of the Book

The book is organized in four chapters.

In Chapter 1 I develop the theory of a continuous-time pure-exchange economy with financial markets and imperfect information. The new feature here is, that imperfect information is treated without assuming that a complete asset market is exogenously given. Instead, an equilibrium with complete markets is endogenously derived. Consumption-based-CAPM-like formulas for the underlying assets are given. Since the underlying assets form a complete market, all other securities are redundant and can be priced by the well-known no-arbitrage arguments.

In the following chapters, the theory developed in Chapter 1 is used to examine the effects of imperfect information and heterogeneity on interest rates. In Chapter 2 I study heterogeneous time preferences among agents taking up the idea of Preferred Habitats developed by Modigliani and Sutch [50]. In contrast to the usual intertemporal models, in which agents uniformly discount the future at some fixed rate $\rho > 0$, I allow for general time preference densities f, letting $f(t)$ describe the weight an agent places on the point in time t. I propose a specific choice for f, the logistic distribution, as a model for preferred habitats. Due to different time preferences, agents trade with each other. It is shown that the more weight an agent places on the future, the more she invests in the asset market in order to consume the proceeds in the future. As a result, preferred habitats may create humps in the yield curve. This effect occurs only in a heterogeneous world in which different habitats are present.

In Chapter 3 a simple model is used to show that imperfect information can have a puzzling impact on the term structure. If all parameters of the model are constant and known, interest rates are deterministic and constant over time and the yield curve is always flat. If one introduces a slight form of imperfect information, assuming that the growth rate of the economy is unknown, then the yield curve is *decreasing*. Moreover, long-term yields converge to the lowest possible value interest rates can have in the model.

Chapter 4 treats the case of different beliefs. The model contains both imperfect information and heterogeneity: Differences in beliefs are modeled as different prior beliefs the agents hold about some unknown factor. As in Chapter 2, I show that in the mixed economy, trade takes place. Bulls borrow money from bears to invest in the risky stock. Interest rates in the heterogeneous world are convex combinations of the interest rates prevailing in economies populated by one kind of investor only. The weights in the convex combination are stochastic. Hence, a wider variety of term structure shapes is possible in the heterogeneous world than in homogeneous worlds. Two examples are given in which the yield curves of the homogeneous and the heterogeneous economies can be computed explicitly. In the first example, I take up the model of Chapter 3. The puzzling result obtained therein is not changed by introducing heterogeneity of beliefs. The yield curves are still decreasing if the agents have different prior beliefs. The second example treats the term structure for an autoregressive specification of the growth rate. The resulting yield curves are convex combinations of the curves obtained in Feldman [24]. Multiple humped yield curves can appear in this case.

To conclude, I restate the main results presented in what follows:

- A financial equilibrium under imperfect information exists. In every such equilibrium, the asset market is complete, although the number of Brownian motions driving the economy is greater than the number of risky assets traded.
- Heterogeneous preferred habitats can explain humps in the yield curve.
- The term structure is decreasing in the usual CCAPM when the constant growth rate is unknown.
- Heterogeneous beliefs lead to higher variations in the yield curve.

1. Imperfect Information and Complete Asset Markets in Continuous Time

1.1 Introduction

This chapter is, first, an introduction to the theory of financial markets in continuous time. The framework is laid out for the applications to the theory of interest rates in the following chapters. Furthermore, I present two extensions of the established theory. The first is of a technical nature, concerned with removing a boundedness assumption made up to now in the pertinent literature, and the second is to demonstrate how imperfect information, in this setting, leads to complete asset markets.

It has been assumed in the proofs of existence of equilibrium with financial markets that the aggregate endowment $(e_t(\omega))$ of the economy is bounded away from zero. This assumption is problematic for a number of reasons. One of them is that in many models, as e. g. in Wang [62] and also in the following Chapters 2 and 3, e is a log-normal process and is, a fortiori, not bounded away from zero. One may always add a positive constant or stop the continuous process (e_t) at the right time, but it is not natural to do so. It is shown herein that the assumption that aggregate endowment is bounded away from zero can be replaced by the weaker assumption that marginal utilities satisfy a certain integrability condition.

Beyond this technical point, an economic theme is also presented - the idea, that, as long as the arrival of new information is described by a Brownian motion, imperfect information leads to complete asset markets.

One of the important allocational aspects of the asset market is, that it allows for the financing of complicated state-contingent consumption/investment plans. An asset market substantially reduces the number of markets needed to reach efficiency. In a continuous-time Brownian setting, an infinite number of contingent forward markets is needed to establish an Arrow-Debreu-equilibrium whereas a finite number of assets is sufficient to establish the same Pareto-optimal allocation of state-contingent consumption streams. However, if the asset market does not allow for insuring against all possible risks, that is, if some factors cannot be hedged, it may, and in general, will be impossible to establish the Pareto-efficient allocation of consumption streams. In this case, the asset market is said to be incomplete.

Imperfect information may help to overcome this deficiency of the market by reducing the number of risks to be hedged. The present model assumes a market that contains more independent Brownian motions than non-numéraire assets. If the agents had perfect information, the available set of assets would not suffice to finance all informationally feasible consumption plans. Hence, the asset market would be incomplete. Here, it is assumed that the information available to the agents is represented by the dividends the assets pay and the underlying Brownian motions are not directly observable. It is shown that in this case the set of available assets suffices to finance all informationally feasible consumption streams. In this sense, the asset market is complete.

I reiterate the purposes of the chapter: the general equilibrium theory with financial markets in continuous time is developed with special emphasis on two new contributions, the removal of the boundedness condition on aggregate endowment, and the proof that imperfect information leads generically to complete asset markets in a Brownian framework.

1.2 A Competitive Financial Market with Imperfect Information

In this section, I present a model of an economy whose dividend payments are subject to a large number of shocks generated by a high-dimensional Brownian motion. Only one risky asset is traded which stands as a representative for an index or the market portfolio. Agents do not observe the Brownian motion directly, but only the dividends paid by the risky asset. In contrast to a large fraction of the finance literature, a representative agent is not assumed right from the start. Instead, agents are allowed to differ in risk aversion, impatience and/or beliefs.

There is one perishable consumption/investment good. The horizon of the economy is $T < \infty$. Uncertainty is modeled by a filtered probability space $(\Omega, \mathcal{A}, P, (\mathcal{F}_t)_{t\geq 0})$ endowed with a k-dimensional Brownian motion W adapted to $\mathcal{F}$. The initial state of the economy is described by a $\mathcal{F}_0$-measurable square-integrable random variable η taking values in $\mathbb{R}^l$. Being $\mathcal{F}_0$-measurable, η is independent of the Brownian motion W.

Aggregate output K of the economy is exogenously given and grows at a rate X, that is

$$\frac{dK_t}{K_t} = dX_t, \quad K_0 = 1\,. \tag{1.1}$$

The evolution of X is described by the stochastic differential equation

$$\left.\begin{aligned} dX_t &= \phi(X,Y)_t dt + \xi(X)_t dW_t, \\ X_0 &= 0\,. \end{aligned}\right\} \tag{1.2}$$

X depends in general also on other factors, Y, described as an $\mathbb{R}^l$-valued stochastic process. Y is modeled as the solution to

$$\left.\begin{aligned} dY_t &= \mu(X,Y)_t dt + \zeta(X,Y)_t dW_t, \\ Y_0 &= \eta\,. \end{aligned}\right\} \tag{1.3}$$

Here, ϕ and μ are $\mathbb{R}$ and $\mathbb{R}^l$-valued, respectively. The diffusion coefficients ξ and ζ take their values in $\mathbb{R}^k$ resp. $\mathbb{R}^{l\times k}$. It is assumed that the functionals ϕ, μ, ξ, and ζ appearing above are nonanticipative, and satisfy a global Lipschitz and linear growth condition. Hence, the stochastic differential equation given by (1.2) and (1.3) has a unique strong solution.

Assumption 1.2.1. *The diffusion coefficient ξ is strongly non-degenerate:*

$$\text{there is an } \epsilon > 0 \text{ with } \quad ||\xi(X)_t|| \geq \epsilon\,. \tag{1.4}$$

Moreover, the following Novikov-type condition is imposed on ϕ: there exists a sequence $0 = t_0 < t_1 < \ldots < t_n = T$ with

$$E \exp\left(\frac{1}{2\epsilon}\int_{t_{m-1}}^{t_m} ||\phi(X,Y)_s||^2\, ds\right) < \infty \quad , 1 \leq m \leq n\,. \tag{1.5}$$

All publicly available information is contained in the growth rate of aggregate output X, the factors Y are unobservable. Denote by $\mathcal{G}_t = \mathcal{F}_t^X \vee \mathcal{N}$ the filtration generated by X, augmented by the sets of zero probability $\mathcal{N}$. Let $\mathcal{O}(\mathcal{G})$ denote the σ-field of all events $A \in \mathcal{A} \otimes \mathcal{B}([0,T])$ which are progressively measurable with respect to the observation filtration $\mathcal{G}$. Henceforth, every event $A \in \mathcal{O}(\mathcal{G})$ and every $\mathcal{O}(\mathcal{G})$-measurable process are called observable.

I agents live in the economy characterized by a felicity function $u^i(t,c)$ and a belief P^i. They choose a rate of consumption $(c_t(\omega))$ out of the consumption space

$$L = \left\{c \in \mathcal{L}^2(\Omega \times [0,T], \mathcal{O}(\mathcal{G}), P \otimes dt) : c \geq 0 \quad P \otimes dt - a.e.\right\}.$$

The felicity function u^i and the belief P^i induce a preference ordering over L represented by a von Neumann-Morgenstern type utility function U^i:

$$U^i(c) = E^i \int_0^T u^i(t, c_t)dt\,. \tag{1.6}$$

Remark 1.2.1. Agents have time-additive von Neumann-Morgenstern utilities. In recent years, this model has been criticized for two reasons. First, in representative agent models, it does not perform well empirically, see Mehra and Prescott for a famous example [45]. Second, it does not distinguish relative risk aversion and elasticity of intertemporal substitution. Several extensions of the standard model have been advanced. Some authors try to reconcile the empirical data with representative agent models by introducing "catching up with the Joneses"-preferences or habit formation preferences (see Abel[1], Constantinides[9], and Detemple and Zapatero[16]). An approach which separates the concepts of risk aversion and intertemporal substitutability is Duffie's and Epstein's [20] stochastic differential utility, which is a stochastic differential formulation of well-known Kreps-Porteus [38] utility. Up to now, however, these preferences have been used in representative agent models only. To my knowledge, there does not exist any model which studies heterogeneous agents with such preferences. This remains, naturally, an interesting task for further research.

Another interesting extension of the standard model has been developed by Hindy, Huang, and Kreps [28]. They focus on the consumption space which, here, consists of the *rates* of consumption. They point out that consumption does not only occur in rates but also in gulps. Mathematically, this means that accumulated consumption is not necessarily absolutely continuous. They then propose the space of right-continuous, increasing processes as a natural choice for the consumption space. From the viewpoint of asset pricing, this is also a natural approach, since stocks pay dividends in rates as well as in lump sums. Within the model of Hindy, Huang, and Kreps, dividend processes belong to the consumption space, which should lead to a more elegant asset pricing theory. Unfortunately, the standard utility (1.6) cannot be extended to a utility function on the space of right-continuous, increasing processes. To obtain a utility function in the Hindy, Huang, Kreps model, one must replace the rate of consumption c_t in (1.6) with some weighted average of past consumption, like $y_t = \int_0^t e^{-\beta(t-s)} c_s ds$. Since my focus is on imperfect information and interest rates rather than the fundamentals of asset pricing, I restrict myself to the classical utility and defer the extension to other utilities to future work.

After this detour, let us continue to set up the model. All agents are risk averse and the felicity functions are smooth:

Assumption 1.2.2. *The felicity functions* $u^i : \mathbb{R}^+ \times \mathbb{R}^+ \to \mathbb{R}$ *are in* $C^{2,3}$; *for all* t, $u^i(t, \cdot)$ *is strictly increasing and strictly concave.*

In addition, I impose the Inada condition on felicity functions to simplify the analysis:

Assumption 1.2.3. *Marginal felicity is infinite at zero:*

$$\lim_{c \to 0} \frac{\partial}{\partial c} u^i(t, c) = \infty .$$

Remark 1.2.2. As I said in the introduction, one aim of the chapter is to remove the assumption that aggregate endowment is bounded away from zero. The assumption has been made in the literature to avoid the difficulties unbounded marginal felicities possibly create. If *all* agents have a felicity function with infinite derivative at zero, marginal felicities, being decreasing functions, remain bounded, regardless of aggregate endowment. In this case, an equlibrium exists as Karatzas, Lehoczky, Shreve [33] and Dana,Pontier [13] show. Therefore, there is only a problem, if some felicity functions have an infinite derivative at zero. I use the stronger assumption that *all* felicities are of this type. However, once this case is understood, it is easy to treat the general case using the arguments developed by Karatzas et al. Assumption 1.2.3 leads to interior solutions for the maximization problems and allows the use of differential methods. I did not want to burden the argument with the technical difficulties a mixture of both types of felicity functions induces, since it is not the essential point in the argument.

The agents may differ in their beliefs about η. η can be interpreted as a structural parameter, which is initially unknown.

To give an example, X could be modeled as a Brownian motion with unknown drift η which agents must learn by observing X (compare Chapters 3 and 4). Heterogeneous prior beliefs may also be viewed as a way to model investors who use different methods of estimation. Herein, the agents have different opinions about the parameter η and obtain, therefore, various estimates of it.

The agents' beliefs are, though, not too far from one another. The agents agree on the events of zero probability; all P^i are mutually equivalent and equivalent to the reference measure P. Moreover, they share the beliefs about the dynamics of the system (X, Y); in particular, W is a standard Brownian motion under all beliefs P^i:

$$\left(P^i\right)^W = P^W .$$

Denote by $g^i(x)$ the density of the law of η under P^i with respect to the law of η under P. Then

$$\frac{dP^i}{dP} = g^i(\eta) ,$$

and I assume that the beliefs about η are not too distinct:

Assumption 1.2.4.

$$g^i(\eta) \in \mathcal{L}^2(\Omega, \mathcal{A}, P) .$$

The assumptions 1.2.2 and 1.2.4 ensure that the utility functions U^i remain finite.

Introduce the density process

$$Z_t^i = E\left[g^i(\eta) \,\middle|\, \mathcal{G}_t\right] = \left.\frac{dP^i}{dP}\right|_{\mathcal{G}_t} . \tag{1.7}$$

Heterogeneity of beliefs is a special case of state-dependent felicity functions as the following lemma shows.

Lemma 1.2.1. *For all consumption streams $c \in L$ and all agents i*

$$U^i(c) < \infty\,.$$

U^i has a generalized von Neumann-Morgenstern representation under the reference measure P with the state-dependent felicity function

$$v^i((\omega,t),d) = u^i(t,d)Z_t^i(\omega)\,, \tag{1.8}$$

that is

$$U^i(c) = E\int_0^T v^i((\omega,t),c_t)\,dt\,.$$

Proof. Let $d \in L$. Since u^i is continuously differentiable in the first and concave in the second variable, there are constants[1] $a,b > 0$ such that for all x

$$u^i(t,x) \le ax + b\,.$$

By assumption 1.2.4 the product $dg^i(\eta)$ is integrable with respect to $P \otimes dt$. Hence,

$$U^i(d) \le aE^i\int_0^T d_t dt + b \le aE\int_0^T d_t g^i(\eta)\,dt + b < \infty\,.$$

By changing the probability measure, one obtains a representation of the utility functions U^i under the global reference measure P:

$$\begin{aligned} U^i(d) &= E^i\int_0^T u^i(t,d_t)dt \\ &= E\int_0^T u^i(t,d_t)Z_t^i\,dt \\ &= E\int_0^T v^i((\cdot,t),d_t)\,dt\,. \end{aligned}$$

Having described the preferences, let us have a look at the endowments. Agent i initially owns a share s^i of the index, $0 \le s^i \le 1$. If she does not trade at all, she receives dividends at a rate $e_t^i = s^i K_t$. e^i is called the endowment of agent i. The

[1] Take, e. g. $a = \max_t \frac{\partial}{\partial c} u^i(t,1)$ and $b = \max_t u^i(t,1)$.

shares held by the agents add up to 1, $\sum_{i=1}^{I} s^i = 1$. Also, a nonzero endowment for agent i, $s^i > 0$ is natural, since an agent who owns nothing initially has no role to play in the economy.

The aggregate endowment is equal to aggregate output, $\sum_i e^i = K$. K is strictly positive because it is equal to

$$K_t = \exp\left(X_t - \frac{1}{2}[X]_t\right).$$

Again, I point out that there need not be a strictly positive lower bound for K. Instead, the restrictive assumption that aggregate endowment is bounded away from zero, is replaced by the following

Assumption 1.2.5. *Marginal felicities of initial endowments are possible Arrow-Debreu prices:*

$$Z_t^i \frac{\partial}{\partial c} u^i(t, e_t^i) \in L.$$

The concept of an Arrow-Debreu price appears here for the first time. An Arrow-Debreu price is a positive, hence continuous, linear functional $\Psi : L \to \mathbb{R}$ assigning a price to every consumption stream $d \in L$. Since every $\mathcal{L}^2$ is a Hilbert space, there is an element ψ, which is in L because of the positivity of Ψ, such that Ψ has the representation $\Psi(d) = E\int_0^T \psi_u d_u \, dt$. Thus, price functionals can be identified with processes $\psi \in L$.

Assumption 1.2.5 cannot be weakened in general. Assume that there is only one agent. Then, to establish a no-trade equilibrium in the Arrow-Debreu sense, it is necessary to find a price $\psi \in L$ that separates the endowment e from the set $G = \{d \in L; U(d) \geq U(e)\}$ of consumption streams preferred to e. The only candidate in a smooth model like this one for such a price process is the marginal felicity $\frac{\partial}{\partial c} u(t, e_t)$. If it is not square-integrable, then there exists no equilibrium. For more on the necessity of Assumption 1.2.5, the reader may consult the overview of Mas-Colell and Zame [44], especially Example 6.5, and the paper of Araujo and Monteiro [2], where it is shown that an equilibrium does generically not exist if one does not have a condition on the integrability of marginal felicities like Assumption 1.2.5.

Moreover, Assumption 1.2.5 implies that the utilities of initial endowments are finite, a fact that will be useful later.

Lemma 1.2.2. *Initial utility is finite for all agents:*

$$U^i(e^i) > -\infty .$$

Proof. By monotonicity, $U^i(e^i) \geq U^i(e^1 \wedge 1)$. For $x \leq 1$, the concavity of v^i yields

$$\begin{aligned} v^i(\cdot, x) &= v^i(\cdot, 1) - \int_x^1 \frac{\partial}{\partial c} v^i(\cdot, s) ds \\ &\geq v^i(1) - (1-x) \frac{\partial}{\partial c} v^i(\cdot, x). \end{aligned}$$

e^i and the marginal utility $\frac{\partial}{\partial c} v^i(\cdot, e^i)$ are in L. Therefore,

$$U^i(e^i) \geq E \int_0^T \left\{ v^i(1) - (1-e^i)^+ \frac{\partial}{\partial c} v^i((t,\omega), e^i_t) \right\} dt > -\infty .$$

Let us have a look at the asset market. The individuals are price takers. The market mechanism produces an observable (ex-dividend) price S for the index. In the sequel, asset prices are taken to be observable Itô-processes. There is a market for lending and borrowing at a short rate r, an observable process. $\beta_t = \exp(\int_0^t r_u du)$ is called the money market account. The prices S and β are determined endogeneously in equilibrium. A spot market for the consumption good opens at every point in time t where the consumption good is traded at a price ψ, taken here to be an Itô-process like S. Using ψ as a deflator, the deflated gain process for the index is

$$G_t = S_t \psi_t + \int_0^t K_u \psi_u du$$

and the deflated value of the money market account is

$$H_t = \beta_t \psi_t .$$

Definition 1.2.1. *The triple of observable processes* (θ^1, θ^2, c) *is called a portfolio/consumption strategy at initial cost* $x \geq 0$ *iff*

$$c \in L \tag{1.9}$$

$$\int_0^T \left|\theta_u^1\right|^2 d[G]_u < \infty \quad a.e. \tag{1.10}$$

$$\int_0^T \left|\theta_u^2\right|^2 d[H]_u < \infty \quad a.e. \tag{1.11}$$

and the deflated value of the portfolio $V_t = (\theta_t^1 S_t + \theta_t^2 \beta_t)\psi_t$ *satisfies*

$$V_t = x + \int_0^t \theta_u^1 dG_u + \int_0^t \theta_u^2 dH_u - \int_0^t c_u \psi_u du \tag{1.12}$$

$$V_t \geq 0 \quad a.e. \quad \text{for all } t \tag{1.13}$$

To avoid the usual change of numéraire argument, the deflated model is used right from the definition of portfolio processes. If (θ^1, θ^2, c) is a portfolio/consumption strategy at cost $\tilde{x} \leq x$, then c is said to be affordable at cost x. Denote by $C(x, S, \beta, \psi)$ the set of all consumption streams affordable at cost x if prices are (S, β, ψ).

Given the initial value S_0 of the index, the agents are endowed with a capital $x^i = s^i S_0$. They choose a consumption strategy that maximizes their utility over all strategies which are affordable at cost x^i. In equilibrium, all markets clear.

Definition 1.2.2. *An Arrow-Radner equilibrium consists of asset prices* (S, β), *a spot price for the consumption good* ψ *and portfolio/consumption strategies* $(\theta^{1i}, \theta^{2i}, c^i)_{i=1,\dots,I}$ *such that agents are rational,*

$$c^i \in \arg\max_{d \in C(x^i, S, \beta, \psi)} U^i(d),$$

and all markets clear:

$$\sum_{i=1}^I \theta_t^{1i} = 1 \quad a.e.$$

$$\sum_{i=1}^{I} \theta_t^{2i} = 0 \quad a.e.$$

$$\sum_{i=1}^{I} c^i = K \quad a.e.$$

1.3 Martingale Representation Theorem for the Innovation Process

Faced with imperfect information, agents engage in Bayesian inference. They estimate the unobservable drift ϕ of X on the basis of the observed realizations $(X_s)_{s\leq t}$ of the rate of aggregate output growth. The difference between the observed growth rate dX_t and the estimated drift $\hat{\phi}$ has the structure of white noise in the sense that there is an *observable* Brownian motion $\hat{W}$ such that

$$dX_t - \hat{\phi}_t dt = \sigma_t d\hat{W}_t \,.$$

In Filtering Theory, the Brownian motion $\hat{W}$ is called the innovation process. The aim of this section is to establish the martingale representation property for the innovation process, that is, every observable local martingale is a stochastic integral with respect to $\hat{W}$. This property is the key to complete markets, as the following sections will demonstrate.

The facts presented in this section are well known in the literature on the statistics of stochastic processes. In this presentation, ideas stemming from Liptser, Shiryaev [39] and Kallianpur [30] have been combined.

Agents estimate the unobservable drift of X as

$$\hat{\phi}_t = E\left[\,\phi(X,Y)_t \,|\, \mathcal{G}_t\right] .$$

The innovation process is given by

$$d\hat{W}_t = \sigma(X)_t^{-1}\left(dX_t - \hat{\phi}_t dt\right) ,$$

with $\sigma(X)_t = ||\xi(X)_t||$.

Lemma 1.3.1. *The innovation process $\hat{W}$ is an observable Brownian motion.*

Proof. By its definition, $\hat{W}$ is observable. Note that $\hat{W}$ can be written as

$$\hat{W}_t = B_t + \int_0^t \frac{\phi(X,Y)_u - \hat{\phi}_u}{\sigma(X)_u} du,$$

with

$$B_t = \int_0^t \frac{\xi(X)_u}{\sigma(X)_u} dW_u.$$

B is a Brownian motion with respect to $\mathcal{F}$ since it is a martingale with quadratic variation $[B]_t = t$. Therefore,

$$E[B_{t+s} - B_t | \mathcal{G}_t] = E[E[B_{t+s} - B_t | \mathcal{F}_t] | \mathcal{G}_t] = 0.$$

Moreover, the law of iterated expectations yields

$$E\left[\int_t^{t+s} \frac{\phi(X,Y)_u - \hat{\phi}_u}{\sigma(X)_u} du \,\middle|\, \mathcal{G}_t\right] = E\left[\int_t^{t+s} \frac{\phi(X,Y)_u - E[\phi(X,Y)_u | \mathcal{G}_u]}{\sigma(X)_u} du \,\middle|\, \mathcal{G}_t\right] = 0.$$

Thus, $\hat{W}$ is a $\mathcal{G}$-martingale. The quadratic variation is

$$\left[\hat{W}\right]_t = [B]_t = t.$$

By Lévy's theorem, $\hat{W}$ is a $\mathcal{G}$-Brownian motion.

With the estimator $\hat{\pi}$ and the innovation process $\hat{W}$ at hand, agents rewrite the dynamics of aggregate output K in observable terms:

$$\frac{dK_t}{K_t} = \sigma_t \hat{\pi}_t dt + \sigma_t d\hat{W}_t. \tag{1.14}$$

They identify, so to speak, an equivalent complete information economy which is given by the filtered probability space $(\Omega, \mathcal{G}_\infty, P, (\mathcal{G}_t))$, the innovation process $\hat{W}$ and the output process K whose dynamics obey (1.14). In the standard model, the filtration is generated by the underlying Brownian motion and the representation theorem holds true. Here, the filtration is not generated by $\hat{W}$, but nevertheless, one has the following result.

Theorem 1.3.1. *Every observable local martingale M admits a representation*

$$M_t = M_0 + \int_0^t \theta_u d\hat{W}_u \tag{1.15}$$

for some observable process θ with

$$\int_0^T \theta_t^2 dt < \infty \quad a.e.$$

In particular, every observable local martingale is continuous.

If M is a martingale, the representation is unique in the sense that for any two processes θ^1, θ^2 satisfying (1.15)

$$\theta^1 = \theta^2 \quad P \otimes dt - a.e.$$

holds.

Proof. A sketch of the proof is as follows. Using the Girsanov theorem, one constructs a measure Q, under which $\bar{X}_t = \hat{W}_t + \int_0^t \frac{\hat{\phi}_u}{\sigma(X)_u} du$ is a Brownian motion. One then shows that $\bar{X}$ generates the same information as X. Therefore, $\mathcal{G}$ is a Brownian filtration. Under Q, one applies the martingale representation theorem for Brownian motion. Finally, it is shown that this yields a representation in terms of $\hat{W}$ for observable P-martingales.

Let us introduce the measure Q under which $\bar{X}$ is a Brownian motion. Set $\pi_t = \frac{\xi(X)_t}{\sigma(X)_t}$ and

$$Z_t = \exp\left(-\int_0^t \pi_u dB_u - \frac{1}{2}\int_0^t \pi_u^2 du\right).$$

For the sequence (t_m) of (1.5) one has

$$E\exp\left(\frac{1}{2}\int_{t_{m-1}}^{t_m} \pi_s^2\, ds\right) \le E\exp\left(\frac{1}{2\epsilon}\int_{t_{m-1}}^{t_m} \|\phi(X,Y)_s\|^2\, ds\right) < \infty$$

by Assumption 1.2.1. By a slight generalization of the Novikov condition (cf. Karatzas, Shreve [34], Theorem 3.5.14), Z is a martingale. Girsanov's theorem states that under the measure Q defined via

$$\left.\frac{dQ}{dP}\right|_{\mathcal{F}_t} = Z_t$$

$\bar{X}$ is a Brownian motion.

Next, we show that indeed $\mathcal{G}$ is generated by the Q-Brownian motion $\bar{X}$.

Lemma 1.3.2. *$\bar{X}$ generates the same information as X:*

$$\mathcal{F}^{\bar{X}} \vee \mathcal{N} = \mathcal{G}.$$

Proof. From its definition, $\bar{X}$ is an observable process, hence

$$\mathcal{F}_t^{\bar{X}} \subset \mathcal{G}_t$$

for all t. On the other hand, X solves the stochastic differential equation

$$dX_t = \sigma(X)_t d\bar{X}_t, \quad X_0 = 0.$$

By a theorem of Yamada, Watanabe [63] (compare Karatzas, Shreve, [34] Corollary 5.3.23., p.310), there is a progressively measurable functional F with $X = F(\bar{X})$ a.e. and therefore $\mathcal{G}_t \subset \mathcal{F}^{\bar{X}} \vee \mathcal{N}$.

The next lemma establishes a representation for the observable density

$$\hat{Z}_t = \left.\frac{dQ}{dP}\right|_{\mathcal{G}_t}.$$

Lemma 1.3.3. *The observable density process $\hat{Z}$ can be written as*

$$\hat{Z}_t = \exp\left(-\int_0^t \hat{\pi}_s d\hat{W}_s - \frac{1}{2}\int_0^t \hat{\pi}_s^2\, ds\right)$$

with $\hat{\pi}_t = E[\pi_t | \mathcal{G}_t]$.

Proof. $\hat{Z}^{-1}$ is a positive, observable Q-martingale. Since $\bar{X}$ is a Brownian motion under Q and by Lemma 1.3.2, $\hat{Z}^{-1}$ is a Brownian martingale. By the well known representation result for Brownian martingales (cf. Karatzas, Shreve [34], Problem 3.4.16), there is an observable process γ with

$$d\hat{Z}_t^{-1} = \gamma_t \hat{Z}_t^{-1} d\bar{X}_t .$$

By Girsanov's theorem,

$$\bar{X}_t - \int_0^t \gamma_u \, du = \hat{W}_t + \int_0^t (\hat{\pi}_u - \gamma_u) du$$

is a Brownian motion under P. Since the innovation process $\hat{W}$ is an observable Brownian motion under P, $\left(\int_0^t (\hat{\pi}_u - \gamma_u) du\right)$ is an observable martingale of bounded variation, thus constant, and $\gamma = \hat{\pi}$ follows. As a stochastic exponential, $\hat{Z}^{-1}$ can be written as

$$\begin{aligned}\hat{Z}_t^{-1} &= \exp\left(\int_0^t \hat{\pi}_s d\bar{X}_s - \frac{1}{2}\int_0^t \hat{\pi}_s^2 \, ds\right) \\ &= \exp\left(\int_0^t \hat{\pi}_s d\hat{W}_s + \frac{1}{2}\int_0^t \hat{\pi}_s^2 \, ds\right) .\end{aligned}$$

To complete the proof of Theorem 1.3.1, let M be an observable local P-martingale and (τ^n) a localizing sequence of stopping times. Then

$$N_t^n := M_{\tau^n \wedge t} \hat{Z}_t^{-1}$$

is a Brownian martingale under Q. By the representation result under Q, there is a unique observable process γ^n with

$$N_t^n = N_0^n + \int_0^t \gamma_u^n d\bar{X}_u .$$

Since γ^n is uniquely determined, it is possible to define $\gamma_t := \gamma_t^n$ on $[0, \tau^n]$. Since $\tau^n \to \infty$,

$$N_t := M_t \hat{Z}_t^{-1} = M_0 + \int_0^t \gamma_u d\bar{X}_u .$$

Now, stochastic integration by parts yields with the aid of Lemma 1.3.3

$$
\begin{aligned}
dM &= \hat{Z}dN + Nd\hat{Z} + d[\hat{Z}, N] \\
&= \hat{Z}\gamma d\bar{X} - \hat{\pi}N\hat{Z}d\hat{W} - \hat{\pi}\gamma\hat{Z}dt \\
&= (\hat{Z}\gamma - M\hat{\pi})d\hat{W} \,.
\end{aligned}
$$

With $\theta = \hat{Z}\gamma - M\hat{\pi}$, this is a representation of M as a stochastic integral over $\hat{W}$.

It remains to show the uniqueness of the representation if M is a martingale. Assume that γ and θ are two observable processes with (1.15). Then $\int(\theta_u - \gamma_u)d\hat{W}_u$ is a martingale identically equal to 0. Hence, the quadratic variation is zero,

$$
\int_0^T (\theta_t - \gamma_t)^2 dt = 0 \,,
$$

thus $\theta = \gamma \quad P \otimes dt - a.e.$

It should be noted that agents can obtain different drift estimates $\hat{\phi}^i$ and different innovation processes $\hat{W}^i$ because they may differ in their beliefs about the unobservable initial state η. The innovation processes $\hat{W}^i$ are related via the formula

$$
\hat{W}_t^i = \hat{W}_t + \int_0^t (\hat{\pi}_u - \hat{\pi}_u^i)\, du \,. \tag{1.16}
$$

The Girsanov theorem yields the following useful representation for the density processes Z^i:

Lemma 1.3.4. *The density processes*

$$
Z_t^i = \left.\frac{dP^i}{dP}\right|_{\mathcal{G}_t}
$$

satisfy the stochastic differential equation

$$
dZ_t^i = (\hat{\pi}_t^i - \hat{\pi}_t) Z_t^i d\hat{W}_t \,. \tag{1.17}
$$

Proof. Apply the representation Theorem 1.3.1 under P to obtain a process γ^1 with

$$
dZ^i = \gamma^1 Z^i d\hat{W} \,.
$$

By Girsanov's theorem,

$$\tilde{W}_t^i = \hat{W}_t - \int_0^t \gamma_u^1 \, du$$

is a P^i-Brownian motion. Apply Theorem 1.3.1 again, but now under P^i, and you get an observable process γ^2 with

$$\tilde{W}_t^i = \int_0^t \gamma_u^2 d\hat{W}_u^i .$$

Equating the two expressions for $\tilde{W}_t^i$, one gets

$$\hat{W}_t - \int_0^t \gamma_u^1 \, du = \int_0^t \gamma_u^2 d\hat{W}_u^i = \int_0^t \gamma_u^2 d\hat{W}_u + \int_0^t \gamma_u^2 (\hat{\pi}_u - \hat{\pi}_u^i) \, du ,$$

or

$$\int_0^t (1 - \gamma_u^2) d\hat{W}_u = \int_0^t \left(\gamma_u^1 + \gamma_u^2 (\hat{\pi}_u - \hat{\pi}_u^i)\right) du .$$

But the left-hand side is a continuous martingale and the right-hand side of bounded variation, hence both sides are zero and $\gamma^2 = 1$ and $\gamma^1 = \hat{\pi}^i - \hat{\pi}$.

1.4 The Existence of an Arrow-Debreu Equilibrium, Pareto Efficiency and the Representative Agent

In the last section, it was demonstrated how the agents deal with the arrival of new information. A second important step is to establish the existence of an Arrow-Debreu equilibrium. If the asset market is complete, a point which shall be demonstrated in the next section, it is possible to implement the allocation of this equilibrium in an Arrow-Radner equilibrium. This argument has been introduced into the literature on continuous-time markets by Duffie and Huang [21]. The idea goes back to Radner [53] and Arrow [3], who proved in a discrete-time setting how financial assets economize on the number of markets needed to reach efficiency.

In general equilibrium models, there are forward markets for all possible states of the world at time 0. Hence, agents can insure completely against all risks at time 0. In the model presented herein, it is equivalent to assume that agents can buy arbitrary consumption streams $c = (c_t(\omega)) \in L$ directly on a market at time 0. They formulate a complete plan of lifetime consumption taking into account all future contingencies and go to the market where they buy their plan directly.

Imperfect information poses no problem to general equilibrium theory. The agents know that they can only demand consumption streams of the type $c_t = c_t((X_s)_{s\le t})$ which depend on the information carried by X only. Thus, the filtration $\mathcal{F}$ one uses for defining the concept of equilibrium given perfect information is replaced with the smaller filtration $\mathcal{G}$. The same argument as in the case of perfect information yields the existence of an equilibrium.

In the general equilibrium framework, $e^i = s^i K$ is the endowment of agent i. Equilibrium is defined thus

Definition 1.4.1. *An Arrow-Debreu-equilibrium consists of a positive linear functional*

$$\Psi : L \to \mathbb{R}$$

and an allocation $(c^i)_{1\le i\le I}$ *such that the consumption market clears,*

$$\sum_i c^i = K\,,$$

and agents maximize utility

$$c^i \in \arg\max U^i(d)$$

over all consumption bundles d*, which are budget feasible:*

$$\Psi(d) \le \Psi(e^i)\,.$$

The classical existence proof of Arrow and Debreu [5] is not valid here because the consumption space is infinite dimensional. Yet, the existence theory for infinite dimensional commodity spaces with a finite number of agents is by now well understood, even though infinite marginal felicity at zero was not covered by the first proofs. For a survey of general equilibrium theory with infinite dimensional commodity spaces, refer to the article by Mas-Colell and Zame [44]. Dana[12] treats the case of additive smooth felicity functions with infinite derivative at zero. Instead of merely citing her results, the important economic ideas underlying the proof, the ideas of Pareto efficiency and a representative agent, are developed to some extent, as these concepts will prove useful in the following chapters.

The method of proof used to establish the existence of an Arrow-Debreu equilibrium was invented by Negishi [51]. It uses the social welfare properties of equilibria to establish the existence of an equilibrium. The first welfare theorem states that, in a competitive equilibrium, the allocation is Pareto efficient. Hence, the set of Pareto efficient allocations is studied first.

Definition 1.4.2. *An allocation $(c^i)_{i=1,\dots,I} \in L^I$ is Pareto efficient iff it is market clearing,*

$$\sum_i c^i = K,$$

and it solves the social welfare problem

$$\sum_i \lambda^i U^i(d^i) = \max_{\sum d^i = K} \tag{1.18}$$

for some weight vector $\lambda \in \Delta^{I-1} = \{z \in \mathbb{R}^I; z^i \geq 0, \sum z^i = 1\}$.

Remark 1.4.1. This definition is slightly uncommon because an allocation is normally called efficient if there is no redistribution of the goods in a way to make everyone at least as well off as before and someone even better off. If, however, an allocation solves the maximization problem (1.18) and $\lambda \gg 0$, then it is clearly efficient in the classical sense. It can also be shown, that for every interior allocation ($c^i \gg 0$) which is efficient in the classical sense, there is a vector of weights λ, so that the allocation solves the social welfare problem (1.18)(cf. Varian [60], ch. 17, p. 334 for this type of argument).

Remark 1.4.2. The set of efficient allocations is described by vectors λ in a compact, convex subset of the finite dimensional space $\mathbb{R}^I$. The typical fixed-point argument will be carried out on this space. By use of this trick, the problems caused by the infinite dimensional commodity space are avoided.

Let us solve the social welfare problem (1.18) for arbitrary aggregate endowments e. Remember that $Z = (Z^1, \ldots, Z^I)$ is the vector of probability densities of the beliefs P^i with respect to the reference measure P, confer (1.7).

Lemma 1.4.1. *Let $\lambda \gg 0$. The value function*

$$\bar{U}(e;\lambda) = \max_{\sum_i d^i = e, d^i \in L} \sum \lambda^i U^i(d^i)$$

has a von Neumann-Morgenstern representation

$$\bar{U}(e;\lambda) = E\int_0^T v(t, e_t, (\lambda^i Z_t^i)_{i=1,\ldots,I})dt\,.$$

The felicity function $v(t,x,\alpha)$ is the solution of

$$v(t,x,\alpha) = \max_{\sum y^i = x} \sum_i \alpha^i u^i(t, y^i)\,. \tag{1.19}$$

v is strictly increasing and strictly concave in x. It is three times continuously differentiable in x and α and twice continuously differentiable in t.

The solutions $y^i(t,x,\alpha)$ of (1.19) are determined by the conditions

$$\alpha^i \frac{\partial}{\partial y} u^i(t, y^i) = \nu \tag{1.20}$$

$$\sum y^i = x \tag{1.21}$$

for a suitable Lagrange multiplier ν. The marginal felicity is

$$\frac{\partial}{\partial x} v(t,x,\alpha) = \nu\,. \tag{1.22}$$

Proof. By the separability of the utilities U^i

$$\sum \lambda^i U^i(d^i) = E\int_0^T \sum(\lambda^i Z_t^i)u^i(t, d_t^i)\, dt$$

and it is clear that it is sufficient to maximize pointwise under the integral.

Fix ω and set $\alpha^i = \lambda^i Z_t^i(\omega)$. By the assumption of infinite marginal utility at zero (1.2.3), the nonnegativity constraint is not binding. Let us solve the Lagrange problem (1.19). (1.20) and (1.21) are the first order conditions for this problem and determine the solutions $y^i(t, x, \alpha)$ and $v(t, x, \alpha)$. The implicit function theorem yields the smoothness in t, x and α required. Now, $v(t, x, \alpha) = \sum \alpha^i u^i(t, y^i(t, x, \alpha))$ is continuously differentiable. By the envelope theorem or direct calculation (1.22) follows.

The function $\bar{U}$ has the properties of a utility function as the preceding lemma shows. Therefore, it is usually interpreted as the utility function of a representative agent.

Now, it is time to introduce the second idea of the Negishi method. This one is related to the second welfare theorem: Every Pareto efficient allocation can be sustained as an equilibrium with transfer payments. It is possible to redistribute wealth among the agents and to define a price Ψ in a manner that Ψ is an equilibrium price for the allocation. If for some allocation, no transfer payments are needed, an equilibrium is found.

Denote by

$$c^i(\lambda)_t = y^i(t, K_t, (\lambda^i Z_t^i)_{i=1,\dots,I}) \tag{1.23}$$

the Pareto efficient allocation associated with λ. The candidate for the price process $\psi(\lambda)$ is the marginal felicity of the representative agent at aggregate endowment. Set

$$\psi(\lambda)_t = \frac{\partial}{\partial x} v(t, K_t, (\lambda^i Z_t^i)_{i=1,\dots,I})\,. \tag{1.24}$$

The money agent i needs in addition to his endowment e^i to buy the consumption stream $c^i(\lambda)$ is given by

$$b^i(\lambda) = E\int_0^T \psi(\lambda)_t(c^i(\lambda)_t - e_t^i)\,dt\,. \tag{1.25}$$

If $b^i(\lambda)$ units of money are given to or taken from agent i, then she is able to afford $c^i(\lambda)$. Note that the sum of payments is $\sum_i b^i(\lambda) = 0$.

Theorem 1.4.1. *Let $\lambda \gg 0$. The price $\Psi(\lambda)$ defined by the price process $\psi(\lambda)$ in (1.24) is an equilibrium price for the allocation $c(\lambda)$ in (1.23) with transfer payments $b(\lambda)$ given by (1.25).*

Proof. The market clears by definition of $c(\lambda)$. The transfer payment $b^i(\lambda)$ ensures that agent i can afford the consumption stream $c^i(\lambda)$.

It remains to show the optimality of $c^i(\lambda)$ for agent i. This is done by a standard Lagrange argument.

Assume that $d \in L$ is budget-feasible for agent i who has obtained the transfer payment $b^i(\lambda)$,

$$\Psi(\lambda)(d) \le \Psi(\lambda)(e^j) + b^i(\lambda) = \Psi(\lambda)(c^i(\lambda))\,.$$

By the definition of $\psi(\lambda)$ (1.24) and the first order condition of the representative agent's problem (1.20), $c^i(\lambda)$ solves the following optimization problem:

$$\max_x \left(v^i((\omega,t),x) - \frac{1}{\lambda^i}\psi(\lambda)_t x\right)\,.$$

Hence,

$$\begin{aligned}
&U^i(d) \le U^i(d) - \tfrac{1}{\lambda^i}\Psi_\lambda(d - c^i(\lambda))\\
&= \tfrac{1}{\lambda^i}\Psi_\lambda(c^i(\lambda)) + E\textstyle\int_0^T v^i((\omega,t),d_t) - \tfrac{1}{\lambda^i}\psi(\lambda)_t d_t\,dt\\
&\le \tfrac{1}{\lambda^i}\Psi(\lambda)(e^i) + E\textstyle\int_0^T \max_x\left(v^i((\omega,t),x) - \tfrac{1}{\lambda^i}\psi(\lambda)_t x\right)dt\\
&= \tfrac{1}{\lambda^i}\Psi_\lambda(c^i(\lambda)) + E\textstyle\int_0^T v^i((\omega,t),c^i(\lambda)_t) - \tfrac{1}{\lambda^i}\psi(\lambda)_t c^i(\lambda)_t\,dt\\
&= U^i(c^i(\lambda))\,.
\end{aligned}$$

The final step consists in looking for a zero of the transfer map b.

Theorem 1.4.2. *There is a strictly positive weight vector $\lambda^* \gg 0$ with $b(\lambda^*) = 0$. In particular, $(\Psi(\lambda^*), c(\lambda^*))$ is an Arrow-Debreu equilibrium.*

Proof. Define the function f on the interior of Δ^{I-1} via

$$f^i = -\frac{1}{\lambda^i} b^i(\lambda) .$$

Every zero of b is a zero of f. Lemma 1.4.3 shows that f has the properties of an excess demand function; especially it satisfies Walras' law

$$\sum_i \lambda^i f^i(\lambda) = \sum b^i(\lambda) = 0 ,$$

because the sum of transfer payments is zero. The classical fixed-point argument using Kakutani's theorem yields a zero for f; confer Mas-Colell, Whinston, Green [43], Chapter 17.C. To apply the fixed-point theorem, one needs to show the continuity of f. A first step in this direction is

Lemma 1.4.2. *The upper envelope of all possible equilibrium prices is square integrable:*

$$\max_\lambda \psi(\lambda) \in L .$$

Proof. The proof of Mas-Colell and Zame [44],Theorem 11.1 is adapted to the case at hand.

The marginal felicity of the representative agent is by (1.22)

$$\frac{\partial}{\partial x} v(t, K_t, (\lambda^i Z_t^i)_i) = \lambda^i Z_t^i \frac{\partial}{\partial x} u^i(t, c^i(\lambda)_t)) \quad a.e.$$

It suffices to check that all marginal utilities

$$Z_t^i \frac{\partial}{\partial x} u^i(t, c^i(\lambda)_t))$$

are square-integrable.

Set $A^i = \{(\omega, t); c^i > e^i\}$, and $A = \cup A^i$. For $(\omega, t) \in A$, say $(\omega, t) \in A^j$, the monotonicity of

$$\frac{\partial}{\partial x} u^j$$

implies in conjunction with the first order condition (1.20)

$$\lambda^i Z_t^i \frac{\partial}{\partial x} u^i(t, c^i(\lambda)_t) = \lambda^j Z_t^j \frac{\partial}{\partial x} u^j(t, c^j(\lambda)_t)$$
$$\leq \lambda^j Z_t^j \frac{\partial}{\partial x} u^j(t, e_t^j).$$

On A^c $c^i \leq e^i$ for all i. But $\sum c^i = K$, hence $c^i = e^i$ on A^c and

$$\lambda^i Z_t^i \frac{\partial}{\partial x} u^i(t, c^i(\lambda)_t) = \lambda^i Z_t^i \frac{\partial}{\partial x} u^i(t, e_t^i).$$

It follows

$$\frac{\partial}{\partial x} v(t, K_t, (\lambda^i Z_t^i)_i) \leq \max_i Z_t^i \frac{\partial}{\partial x} u^i(\cdot, e_t^i).$$

The right-hand side is in L by Assumption 1.2.5.

The following lemma establishes the properties of f which allow to conclude that a zero of f, and thus an equilibrium, exists.

Lemma 1.4.3. *The map f has the following properties:*

1. *f is continuous in $\lambda \gg 0$,*
2. *Walras' law holds: for all $\lambda \gg 0$ $\sum_i \lambda^i \epsilon^i(\lambda) = 0$,*
3. *f is bounded below,*
4. *Whenever $\lambda^n \to \lambda \neq 0$ and $\lambda^i = 0$ for some i, then*

$$\max_i f^i(\lambda^n) \to \infty.$$

Proof. 1. By Lemma 1.4.1, the price process $\psi(\lambda)$ and the allocation $c(\lambda)$ are, for fixed (t, ω), continuous in λ. Hence,

$$\psi(\lambda^n)(e^i - c^i(\lambda^n)) \to \psi(\lambda)(e^i - c^i(\lambda)) \qquad P \otimes dt - a.e.$$

Using the preceding Lemma 1.4.2

$$\left| \psi(\lambda^n)(e^i - c^i(\lambda^n)) \right| \leq (\max_\lambda \psi(\lambda)) K \in \mathcal{L}^1.$$

By the theorem of dominated convergence, f is continuous.

2. Walras' law holds by definition of f.

3. The concavity of v^i implies for all x, y

$$\frac{\partial}{\partial x} v^i(\cdot, x)(y - x) \geq v^i(y) - v_i(x) .$$

Because of the first-order condition (1.20)

$$\begin{aligned} f^i(\lambda) &= E \int_0^T \frac{1}{\lambda^i} \psi(\lambda)_t (e_t^i - c^i(\lambda)_t)\, dt \\ &\geq U(e^i) - U(c^i(\lambda)) \\ &\geq U(e^i) - U(K) . \end{aligned}$$

The last difference is a lower bound for f. It is finite because of Lemma 1.2.2.

4. Assume $\lambda^n \to \lambda \neq 0$ and without loss $\lambda_1 = 0$ and $\lambda_2 > 0$. It suffices to show $\liminf \lambda_1^n f^1(\lambda^n) > 0$.
The first order condition (1.20) yields

$$\frac{\lambda_1^n}{\lambda_2^n} = \frac{\frac{\partial}{\partial x} v^2\left((\omega, t), c^2(\lambda^n)_t\right)}{\frac{\partial}{\partial x} v^1\left((\omega, t), c^1(\lambda^n)_t\right)} \geq \frac{\frac{\partial}{\partial x} v^2\left((\omega, t), K_t\right)}{\frac{\partial}{\partial x} v^1\left((\omega, t), c^1(\lambda^n)_t\right)} .$$

The left-hand side tends to zero, hence $c^1(\lambda^n)_t \to 0$, too. Whenever the weight of an agent vanishes, his consumption vanishes, too. One obtains

$$\begin{aligned} \lambda_1^n f^1(\lambda^n) &= E \int_0^T \psi(\lambda^n)_t (e_t^1 - c^1(\lambda^n))\, dt \\ &= E \int_0^T \lambda_2^n \frac{\partial}{\partial x} v^2((\omega, t), c^2(\lambda^n)_t)(e_t^1 - c^1(\lambda^n)_t)\, dt \\ &\geq E \int_0^T \lambda_2^n \frac{\partial}{\partial x} v^2\left((\omega, t), K_t\right) (e_t^1 - c^1(\lambda^n)_t)\, dt \\ &\to \lambda_2 E \int_0^T \frac{\partial}{\partial x} v^2\left((\omega, t), K_t\right) e_t^1\, dt > 0 \end{aligned}$$

where dominated convergence is used again.

1.5 Completeness of the Market and Existence of a Financial Equilibrium

The Arrow-Debreu equilibrium derived in the last section characterizes the efficient allocation $c(\lambda)$ that is achieved if a complete set of forward markets exists at time 0. Now let us see whether it is possible to reach the same allocation if trade takes place on the spot market for the consumption good and the asset market given by the index and the money account only. The agents want to consume the stream $c^i(\lambda)$. Since they cannot buy the whole stream directly, they have to form a portfolio (θ^1, θ^2) that finances the stream $c^i(\lambda)$ at the cost of their initial wealth $x^i = s^i S_0$. Does such a portfolio exist? The answer is yes if the asset market is complete.

Definition 1.5.1. *Let the prices (S, β, ψ) be given. The asset market is called complete iff for all $c \in L$ there exists a portfolio (θ^1, θ^2) that finances c at some initial cost x, formally*

$$L = \cup_{x \geq 0} C(x, S, \beta, \psi).$$

In the following, the Arrow-Debreu price process $\psi = \psi(\lambda)$ is used as the spot price for the consumption good. Note that by (1.24)

$$\psi(\lambda)_t = \frac{\partial}{\partial x} v(t, K_t, (\lambda^i Z_t^i)_{i=1,\dots,I}) \tag{1.26}$$

and according to Lemma 1.4.1 ψ is a smooth function of Itô-processes, hence itself an Itô-process with some representation $d\psi = \alpha^1 dt + \alpha^2 d\hat{W}$.

If ψ is used as a deflator, $\beta\psi$ is a (local) martingale. Having this in mind, define the short rate r and the market price of risk p as

$$r = -\frac{\alpha^1}{\psi} \tag{1.27}$$

and

$$p = -\frac{\alpha^2}{\psi},$$

hence

$$\frac{d\psi_t}{\psi_t} = -r_t dt - p_t d\hat{W}_t. \tag{1.28}$$

Note that ψ is continuous and strictly positive. For almost all ω, the path (ψ_t) is bounded away from zero. Therefore, the integrals $\int_0^T |r_u|\, du$ and $\int_0^T p_u^2 du$ exist almost surely. Thus,

$$H_t = \exp\left(-\int_0^t p_u d\hat{W}_u - \frac{1}{2}\int_0^t p_u^2\, du\right)$$

and $\beta_t = \exp\left(\int_0^t r_u du\right)$ are well defined. Note that H is the deflated value of the money market account β because of

$$\psi\beta = H.$$

Define the ex-dividend price of the index as

$$S_t\psi_t := E\left[\int_t^T K_u\psi_u\, du \,\middle|\, \mathcal{G}_t\right]. \tag{1.29}$$

The associated deflated gain process G is the observable martingale

$$G_t = E\left[\int_0^T K_u\psi_u\, du \,\middle|\, \mathcal{G}_t\right].$$

As a positive martingale, G has a representation

$$G_t = G_0 - \int_0^t \gamma_u G_u d\hat{W}_u \tag{1.30}$$

by Theorem 1.3.1.

The two assets give rise to two (local) martingales and there is one Brownian motion $\hat{W}$ that generates the space of martingales. The usual argument along the lines of Black-Scholes [8] allows to establish the completeness of the market as long as the volatilities satisfy a certain condition.

Theorem 1.5.1. *If*

$$pS\psi \neq \gamma G \quad P \otimes dt - a.e. \tag{1.31}$$

holds, then the asset market is complete. The initial capital needed to finance a consumption stream $c \in L$ is given by the Arrow-Debreu price $\Psi(c) = E \int_0^T \psi_t c_t \, dt$.

Proof. Let $c \in L$ be given. The martingale

$$M_t = E\left[\int_0^T c_t \psi_t dt \,\middle|\, \mathcal{G}_t\right] \tag{1.32}$$

is a stochastic integral by Theorem 1.3.1:

$$M_t = \Psi(c) + \int_0^t m_u d\hat{W}_u \,. \tag{1.33}$$

Set

$$V_t = E\left[\int_t^T c_u \psi_u \, du \,\middle|\, \mathcal{G}_t\right] \tag{1.34}$$

For (t, ω) fixed, let θ^1, θ^2 be the solution to the linear system

$$-\theta^1 \gamma G - \theta^2 pH = m \tag{1.35}$$

$$\theta^1 S\psi + \theta^2 H = V \,. \tag{1.36}$$

The system has almost always a solution because the determinant is

$$H(-\gamma G + pS\psi) \neq 0$$

by Assumption (1.31). V is the deflated value of the portfolio formed by (θ^1, θ^2) because of (1.36). By (1.34), V is nonnegative and equal to

$$V_t = M_t - \int_0^t c_u \psi_u du \,,$$

by (1.32). (1.33) yields

$$V_t = \Psi(c) + \int_0^t m_u d\hat{W}_u - \int_0^t c_u \psi_u du .$$

Use (1.35) to obtain

$$\begin{aligned} V_t &= \Psi(c) - \int_0^t (\theta_u^1 \gamma_u G_u + \theta_u^2 p_u H_u) d\hat{W}_u - \int_0^t c_u \psi_u du \\ &= \Psi(c) + \int_0^t \theta_u^1 dG_u + \int_0^t \theta_u^2 dH_u - \int_0^t c_u \psi_u du . \end{aligned}$$

Hence, (θ^1, θ^2) finances c at initial cost $\Psi(c)$.

With complete asset markets, it is possible to finance all consumption streams by trading in the asset market. It is no wonder then, that the Pareto efficient allocation of the Arrow-Debreu equilibrium is also the allocation of a financial equilibrium. Denote by

$$\tilde{C}(e, \Psi) = \{c \in L; \Psi(c) \leq \Psi(e)\}$$

the budget set of an agent with endowment e in the Arrow-Debreu framework where Ψ is the price functional associated with ψ.

Lemma 1.5.1. *The budget sets in the Arrow-Debreu- and the Arrow-Radner framework coincide:*

$$C(x^i, S, \beta, \psi) = \tilde{C}(e^i, \Psi) .$$

Proof. Remember that the initial capital is the same in both cases:

$$x^i = s^i S_0 = s^i E \int_0^T K_t \psi_t \, dt = \Psi(e^i) .$$

If c is budget-feasible in the Arrow-Debreu sense, $c \in \tilde{C}(e^i, \Psi)$, then, by Theorem 1.5.1, there exists a portfolio (θ^1, θ^2) that finances c at initial cost $\Psi(c) \leq \Psi(e^i) = x^i$. Hence, $c \in C(x^i, S, \beta, \psi)$.

If, on the other hand, (θ^1, θ^2, c) is a portfolio at initial cost $x \leq x^i$, then the sum of the deflated value V of the portfolio and cumulated consumption is

$$V_t + \int_0^t c_u \psi_u du = x + \int_0^t \theta_u^1 dG_u + \int_0^t \theta^2 dH_u .$$

Because of $V \geq 0$, the right-hand side is a nonnegative local martingale, hence a supermartingale. It follows

$$\Psi(c) = E\int_0^T c_t \psi_t dt \leq E\left(V_T + \int_0^T c_u \psi_u du\right) \leq x$$

and $c \in \tilde{C}(e^i, \Psi)$.

The Arrow-Radner equilibrium cannot be far away - and here it is:

Theorem 1.5.2. *Under Condition (1.31), the allocation* $c(\lambda^*)$ *of the Arrow-Debreu equilibrium derived in Theorem 1.4.2 and asset prices* S *and* β *as defined in (1.29) and (1.27) form an Arrow-Radner equilibrium.*

Proof. Choose portfolio strategies $(\theta^{1i}, \theta^{2i})$ that finance c^i for agent i at cost $x^i = s^i S_0$ as in the proof of Theorem 1.5.1. In particular, the deflated value of i's portfolio is

$$V_t^i = E\left[\int_t^T c_u^i \psi_u du \,\middle|\, \mathcal{G}_t\right] . \tag{1.37}$$

Since the budget sets coincide by Lemma 1.5.1, c^i maximizes utility over all affordable consumption streams for agent i. The consumption market clears because c is the allocation of an Arrow-Debreu equilibrium. It remains to show that the asset market clears. Set $\theta^1 = \sum_i \theta^{1i}, \theta^2 = \sum_i \theta^{2i}$ and $V = \sum_i V^i$. Then

$$\begin{aligned}
V_t &= \sum_i \left(s^i S_0 + \int_0^t \theta_u^{1i} dG_u + \int_0^t \theta_u^{2i} dH_u - \int_0^t c_u^i \psi_u du\right) \\
&= S_0 + \int_0^t \theta_u^1 dG_u + \int_0^t \theta_u^2 dH_u - \int_0^t K_u \psi_u du \\
&= S_0 + G_t - G_0 + \int_0^t (\theta_u^1 - 1) dG_u + \int_0^t \theta_u^2 dH_u - \int_0^t K_u \psi_u du \\
&= S_t \psi_t + \int_0^t (\theta_u^1 - 1) dG_u + \int_0^t \theta_u^2 dH_u .
\end{aligned}$$

On the other hand, (1.37) and the clearing of the consumption market imply

$$V_t = E\left[\left[\int_t^T K_u \psi_u du \middle| \mathcal{G}_t\right]\right] = S_t \psi_t . \tag{1.38}$$

Hence, for all t

$$\begin{aligned} 0 &= \int_0^t (\theta_u^1 - 1) dG_u + \int_0^t \theta_u^2 dH_u \\ &= \int_0^t \left\{ (\theta_u^1 - 1) g_u S_u \psi_u - \theta_u^2 p_u H_u \right\} d\hat{W}_u . \end{aligned}$$

Therefore, the diffusion coefficient vanishes:

$$(\theta_u^1 - 1) g_u S_u \psi_u - \theta_u^2 p_u H_u = 0 . \tag{1.39}$$

Moreover, one has

$$\theta_u^1 S_u \psi_u + \theta_u^2 H_u = S_u \psi_u . \tag{1.40}$$

The unique solution of the linear system (1.39),(1.40) is $\theta^1 = 1, \theta^2 = 0$. The asset market clears.

1.6 Pricing Redundant Securities and the Term Structure of Interest Rates

The equilibrium derived in the previous section provides us with prices for the stock and for the money market account. What can one say about other securities, such as zero-coupon bonds? In this section I show that, since the asset market is complete[2], any security other than the stock and the money market account is redundant, in the sense that it can be duplicated by a portfolio in said stock and money market account. Therefore, the market has a price for all securities. In particular, the term structure of interest rates is determined.

---

[2] if (1.31) holds, what I assume in this section.

Theorem 1.5.1 shows how an agent can finance an arbitrary consumption stream $c \in L$ by some portfolio (θ^1, θ^2) in the stock and the money market account. If one gives a slightly different twist to this statement, one sees immediately how to price an arbitrary security. Replacing the term 'consumption stream' with 'dividend', let us consider the case where some security yields a dividend at rate $d \in L$. Then, by the same Theorem 1.5.1, the security's dividend can be duplicated by a portfolio in the stock and the money market account at some initial cost $S_0^d = \Psi(d) = E\int_0^T d_t \psi_t \, dt$. The capital needed at time t to duplicate the dividend of the security from time t up to time T is

$$S_t^d = \frac{1}{\psi_t} E\left[\int_t^T d_u \psi_u \, du \,\middle|\, \mathcal{G}_t \right] .$$

Since the aim is also to price securities which pay lump-sums, the following theorem allows for somewhat more general processes of cumulative consumption:

Theorem 1.6.1. *Let C be an increasing observable process with*

$$E \int_0^T \psi_u dC_u < \infty .$$

Then there exists a portfolio (θ^1, θ^2) which duplicates C in the sense that the deflated value of the portfolio

$$S_t^C \psi_t = \left(\theta_t^1 S_t + \theta_t^2 \beta_t \right) \psi_t$$

satisfies

$$d(S_t^C \psi_t) = \theta_t^1 dG_t + \theta_t^2 dH_t - \psi_t dC_t . \qquad (1.41)$$

Proof. The proof is the same as for Theorem 1.5.1. One just starts with the martingale

$$M_t = E\left[\int_0^T \psi_u dC_u \,\middle|\, \mathcal{G}_t \right]$$

instead of (1.32) and repeats the same arguments.

I interpret the value S^C of the portfolio duplicating the payoff C as the equilibrium price for the security. It can be shown that any other price process would lead to arbitrage opportunities. In equilibrium, however, there cannot exist any arbitrage opportunity or free lunch because the agents, facing no budget-constraint any more, would consume arbitrary large amounts of the consumption good. In this sense, the price process S^C is the only one consistent with the equilibrium.

It is well known since the work of Harrison, Kreps [25] and Harrison, Pliska [26] that the absence of arbitrage opportunities is (cum grano salis) equivalent to the existence of an equivalent martingale measure. A measure $P^\star$ equivalent to the reference measure P is called an equivalent martingale measure if the sum of the discounted value of the stock and cumulated dividend payments

$$\frac{S_t}{\beta_t} + \int_0^t \frac{K_u}{\beta_u}\,du$$

is a martingale under $P^\star$. Under suitable integrability conditions the existence of a price functional is also equivalent to the existence of an equivalent martingale measure as the next theorem shows(compare also Duffie [19], Section 6.J. and Karatzas [31], Chapter 3, who both have stronger boundedness assumptions).

Theorem 1.6.2. *Assume*

$$EH_T = 1\,. \tag{1.42}$$

Then there exists an equivalent martingale measure $P^\star$ whose density with respect to P on $\mathcal{G}_T$ is

$$\frac{dP^\star}{dP} = H_T\,. \tag{1.43}$$

Proof. $P^\star$ is well defined by (1.43) because of $EH_T = 1$. But this integrability implies also that the supermartingale H has constant expectation; therefore, H is a martingale and it follows that the density process is given by H:

$$\left.\frac{dP^\star}{dP}\right|_{\mathcal{G}_t} = H_t \,.$$

It is just an application of Bayes' rule to the pricing formula (1.29) to show that $P^\star$ is an equivalent martingale measure:

$$\begin{aligned}\frac{S_t}{\beta_t} &= \frac{1}{\beta_t \psi_t} E\left[\int_t^T K_u \psi_u \, du \,\middle|\, \mathcal{G}_t\right] \\ &= \frac{1}{H_t} E\left[\int_t^T K_u \beta_u^{-1} H_u \, du \,\middle|\, \mathcal{G}_t\right] \\ &= E^\star \left[\int_t^T K_u \beta_u^{-1} \, du \,\middle|\, \mathcal{G}_t\right] .\end{aligned}$$

At the end of this chapter, let us have a look at the interest rates in equilibrium. In the last section, it turned out that the short rate can be identified as minus the expected rate of growth of the marginal felicity of a representative agent. It is therefore possible to describe the short rate in terms of partial derivatives of the felicity function v and the drift and diffusion coefficients of the aggregate endowment K and the densities Z^i (see lemma 1.6.1 below). Moreover, Theorem 1.6.1 shows how to price zero-coupon bonds. Thus, the term structure of interest rates $(y_t^{t+\tau})_{0 \le \tau \le T-t}$, where $y_t^{t+\tau}$ is the effective interest rate at time t of a loan with time to maturity τ, is determined[3].

Theorem 1.6.3. *The equilibrium price of a zero-coupon bond with maturity $T' \le T$ is*

$$B_t^{T'} = E\left[\frac{\psi_{T'}}{\psi_t} \,\middle|\, \mathcal{G}_t\right] \tag{1.44}$$

$$= E\left[\frac{\frac{\partial}{\partial x} v(T', K_{T'}, (\lambda^i Z^i_{T'})_{i=1,\dots,I})}{\frac{\partial}{\partial x} v(t, K_t, (\lambda^i Z^i_t)_{i=1,\dots,I})} \,\middle|\, \mathcal{G}_t\right] . \tag{1.45}$$

[3] $(y_t^{t+\tau})$ is the yield of a zero-coupon bond with maturity $t + \tau$. By definition, the interest rate and the price of the zero-coupon bond $B_t^{t+\tau}$ are related via the formula

$$B_t^{t+\tau} = \exp\left(-\tau y_t^{t+\tau}\right) .$$

Proof. Immediately from Theorem 1.6.1 with $C_t = 1_{\{t \geq T'\}}$.

The price process ψ determines the term structure of interest rates. The equation (1.44) will be the starting point for our investigations on the theory of interest rates in the following chapters.

As I said in the beginning, it is possible to study the relationship between the partial derivatives of the felicity of the representative agent and the short rate.

Lemma 1.6.1. *Set* $Z(\lambda)_t = (\lambda^i Z_t^i)_{i=1,\dots,I}$. *The short rate* r *satisfies*

$$\begin{aligned}
r_t = &-\frac{v_{xt}}{v_x}(t, K_t, Z(\lambda)_t) \\
&- \frac{v_{xx}}{v_x}(t, K_t, Z(\lambda)_t)\hat{\pi}_t \sigma_t K_t \\
&- \frac{1}{2}\frac{v_{xxx}}{v_x}(t, K_t, Z(\lambda)_t)\sigma_t^2 K_t^2 \\
&- \sum_i \frac{v_{xx\alpha^i}}{v_x}(t, K_t, Z(\lambda)_t)\lambda^i Z_t^i(\hat{\pi}_t^i - \hat{\pi}_t)\sigma_t K_t \\
&- \sum_{i<j} \frac{v_{x\alpha^i\alpha^j}}{v_x}(t, K_t, Z(\lambda)_t)\lambda^i Z_t^i(\hat{\pi}_t^i - \hat{\pi}_t)\lambda^j Z_t^j(\hat{\pi}_t^j - \hat{\pi}_t)
\end{aligned} \tag{1.46}$$

where subscripts denote partial derivatives.

Proof. This is an application of Itô's lemma and the stochastic differential equations for (Z^i) and K obtained in (1.17) and (1.14).

Several observations can be made. First, the short rate depends on the *rate of time preference*

$$-\frac{v_{xt}}{v_x}$$

of the representative agent. The more impatient the agents are, the higher the prevailing short rate.

The second summand is the product of relative risk aversion

$$-\frac{v_{xx}}{v_x}x$$

and the observable drift of the growth rate $\sigma\hat{\pi}$. A faster growth of the economy tends to increase the short rate. The same, can be argued, should be true for risk aversion because a higher degree of risk aversion should lead to a higher demand for the locally riskless asset. However, risk aversion can also have a lowering effect on the short rate. This is reflected in the third term which is the product of relative risk aversion, relative prudence[4]

$$-\frac{v_{xxx}}{v_{xx}}x,$$

and the volatility σ^2 of the growth rate.

The last two summands in (1.46) are brought about by the heterogeneity of beliefs among the agents. The term

$$-\sum_i \frac{v_{xx\alpha^i}}{v_x}(t, K_t, Z(\lambda)_t)\lambda^i Z_t^i(\hat{\pi}_t^i - \hat{\pi}_t)\sigma_t K_t$$

merely adjusts the short rate for the choice of the reference measure. If all agents share the same belief P^1, which by coincidentally is not equal to the reference measure P, then this term changes the short rate in a manner so that the (irrelevant) estimate $\hat{\pi}$ under P is replaced with the estimate $\hat{\pi}^1$ under the common belief P^1.

Finally, the last term is a quadratic form in, what may be called, the *weighted vector of dispersion of beliefs*

$$(\lambda^i Z^i(\hat{\pi}^i - \hat{\pi}))_i.$$

If the marginal felicity of the representative agent is convex[5] in the vector of weights α, then the matrix

$$\left(\frac{v_{x\alpha^i\alpha^j}}{v_x}(t, K_t, Z(\lambda)_t)\right)_{ij}$$

is positive definite and the last term is negative.

[4] The concept of *prudence* has been introduced by Kimball [36] in analogy to the concept of Arrow-Pratt [4][52] risk aversion. It measures the intensity of the precautionary savings motive in intertemporal decision problems.

[5] But it need not be convex, in general.

2. Heterogeneous Time Preferences - The Preferred Habitat Theory Revisited

As a first application of the general theory to the theory of the term structure, I study the Preferred Habitat Theory advanced by Modigliani and Sutch [50]. Modigliani and Sutch were primarily concerned with an empirical investigation of the effects of a particular policy of the US government and Federal Reserve, "Operation Twist", which aimed at "twisting" the yield curve by raising the yield of bonds with short term to maturity and lowering simultaneously long-term rates. It was believed that this policy would help to reduce the balance-of-payments problem of the US. In order to give a rationale for the linear model they estimated, they reviewed the existing theories of the term structure, mainly the Expectations and the Liquidity Premium Hypothesis, and they developed their own theory, which has become known under the name of 'Preferred Habitat Theory'.

The Liquidity Premium Hypothesis , first formulated by Hicks [27], builds on the idea that the presupposed riskiness of a long-term bond induces investors to require a premium for holding these assets. A long-term bond, though, is certainly riskless at maturity, because its final value is certain. Modigliani and Sutch pointed out that, if an investor's investment horizon coincides with the maturity of a zero-coupon bond and the investor is only interested in the final payoff of the portfolio, risk averse investors will consider the long-term bond as riskless and require a premium for holding short-term bonds. In their words:

> The Hicksian model assumes that all traders are concerned with the short period return and that, therefore,

> anybody going long is bearing the risk associated with the uncertainty of the short period return from longer-term instruments. But this view would be correct only if we could assume that every lender desires to turn his portfolio back into cash at the end of the short period; i.e. that he has a short habitat. In reality, however, different investors are likely to have different habitats ...

and for those with a long-term habitat, it is risky to roll over the money at the stochastic short rate.

The conclusion I draw from this reflection is that one should try to model "preferred habitats" within an intertemporal model and study the impact on the term structure. Therefore, in Section 2.1 a new class of intertemporal preferences which display preferred habitats is introduced.

New preferences, though, are not the only point[1], heterogeneity plays also an important part in this chapter. If all agents have the same habitat, there are no gains from trade and, as will be seen, there is essentially no impact of the habitat on interest rates. If, however, distinct habitats are present in the economy, agents have an incentive to trade, and interesting shapes of the term structure arise.

2.1 Modeling Preferred Habitat Time Preferences

The purpose of this section is to model intertemporal utility functions which describe people with a higher propensity to consume at a certain point in time H, their habitat, than at other times. Think of a person who invests in the bond market with the specific goal of insuring herself against a decreasing standard of living upon retiring. Such an investor has a higher preference for consuming out of the portfolio when retiring than at the time the investment is made. She has a long habitat.

---

[1] and if they were, one could argue that I write the solution of the problem into the preferences. I hope the reader will finally agree with me that this is not the case.

The typical intertemporal utility function over contingent consumption streams $(c_t(\omega))$ takes the form

$$U(c) = E\int_0^\infty e^{-\rho t}u(c_t)\,dt$$

for some *rate of time preference* $\rho > 0$. For the heuristic discussion of this section, the horizon is set equal to ∞.

This utility function is characterized by three properties: independence, stationarity and impatience. The utility function U satisfies the axiom of cardinal independence for states of nature as well as points in time. If two processes c and d coincide on an observable set $A \subset \Omega \times [0, \infty[$ and c is preferred to d,

$$U(c) > U(d),$$

then one can arbitrarily alter the common value of c and d on A without changing the preference ordering: for some arbitrary c' and $\tilde{c} = c'1_A + c1_{A^c}$ and $\tilde{d} = c'1_A + d1_{A^c}$,

$$U(\tilde{c}) > U(\tilde{d})$$

holds true.

Since I will keep the assumption of independence, only the intertemporal aspect of the utility function U will be considered. For a deterministic consumption stream (c_t), the utility is

$$U(c) = \int_0^\infty e^{-\rho s}u(c_s)\,ds\,.$$

The preference ordering induced by U is stationary in the sense that the delayed utility U_t given by

$$U_t(c) = \int_t^\infty e^{-\rho s}u(c_s)ds$$

leads to the same ordering of (delayed) consumption streams:

$$U_t(c) = e^{-\rho t}U((c_{t+s})_{s\geq 0})\,.$$

Endowed with a certain amount of money, the investor with such an intertemporal utility function prefers the same type of consumption streams most at any time. In particular, the rate of time preference ρ remains constant over time.

The last important property of U is that of impatience, $\rho > 0$. A positive rate of time preference implies that the long distant future does not matter much for current decisions. This is frequently used as a justification for the assumption of impatience.

My claim is that people with a (long) habitat do not have stationary time preferences nor are they globally impatient. If one has a habitat, for example, of ten years, then it is plausible to assume that one is patient up to said habitat and that one becomes very impatient at and after the habitat. As a consequence, one's rate of time preference changes over time. Thus, the preferences are not stationary.

To model preferred habitat preferences, I drop the assumption of stationarity and (global) impatience. I keep, however, the property that the very long distant future does not matter much for current decisions by assuming that asymptotically, the time preference density f decreases exponentially: there is a $\bar{\rho} > 0$ with

$$f(t) \sim e^{-\bar{\rho}t} \qquad (t \to \infty) .$$

Strictly positive and smooth (C^2) functions f with finite integral,

$$\int_0^\infty f(t)dt < \infty ,$$

are henceforth called time preference densities.

Definition 2.1.1. *A time preference density f is said to display a preferred habitat in $h > 0$ iff*

- *f has a unique maximum in h;*
- *f decreases exponentially in the long run, that is, there is a long-run rate of time preference $\bar{\rho} > 0$ with*

$$\lim_{t \to \infty} f(t)e^{\bar{\rho}t} = 1 .$$

Note that the standard time preference density $e^{-\rho t}$ is included in this definition. It displays a habitat at $h = 0$ and describes the short-run interested individuals of the Hicksian world.

There is a class of time preference densities that is both easy to handle and displays preferred habitats at times $h > 0$, namely the class of logistic densities,

$$f(t;h,\gamma) = \gamma \frac{\exp(-\gamma(t-h))}{(1+\exp(-\gamma(t-h)))^2},$$

see Figure 2.1. The corresponding distribution functions are

$$F(t;h,\gamma) = (1+\exp(-\gamma(t-h)))^{-1}.$$

h is the unique maximum, the habitat, and γ inversely relates to the variance of F, which is $\pi^2/3\gamma^2$.

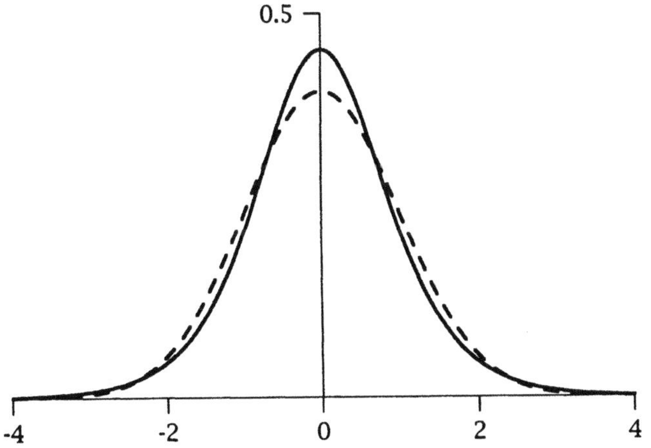

Fig. 2.1. Density of the logistic distribution with mean 0 and variance 1. The dashed line represents the density of the standard normal distribution.

The logistic densities will be the reference model for preferred habitat preferences and will be used to illustrate the results. I wish to stress that the structural results are independent of the specific form chosen for f.

Before concluding, a final remark is in order. It is important not to confuse stationarity and time consistency. Individuals are said to act in a time inconsistent manner, if they wish to revise the plan chosen at time 0 at a later time $t > 0$[2]. Here, individuals' rates of time preference change over time. This change, though, is foreseen, and an adequate consumption plan is formulated in such a way as to avoid the necessity of a later revision. Their preferences are not stationary, but they act in a time consistent manner.

2.2 A Model with Heterogeneous Time Preferences

A model in which agents have distinct time preferences is presented in this section. Since the focus is on time preferences, the remainder of the model is taken to be as simple as possible but as general as necessary. The notation is as in Chapter 1.

The Brownian motion W is one-dimensional. There is perfect information, the factors Y play no role (assume that Y is deterministic and known, e.g.). The observation filtration is equal to the underlying filtration

$$\mathcal{G} = \mathcal{F} .$$

The output of the economy, or aggregate endowment, is

$$\frac{dK_t}{K_t} = dX_t ,$$

and the growth rate X is modeled as the solution of the stochastic differential equation

$$dX_t = (\phi_1 - \phi_2 X_t)dt + \xi dW_t ,$$

with constant parameters $\phi_1, \phi_2, \xi \in \mathbb{R}$. Two distinct cases are treated. For $\phi_2 = 0$, X is a Brownian motion with drift ϕ_1, and

[2] Time preferences of the type "always discount the *next* period higher than the period following the next period" lead to time inconsistent behavior. Such time preferences were first studied by Strotz [57].

for $\phi_2 > 0$ X is a stationary Ornstein-Uhlenbeck process. It is useful to have the conditional distributions of the increments of X at hand. In the first case, $\phi_2 = 0$, these are

$$\mathcal{L}(X_{t+s} - X_t | \mathcal{F}_t) = N(\phi_1 s, \xi^2 s) . \tag{2.1}$$

For $\phi_2 > 0$, X is explicitly given by

$$X_{t+s} = e^{-\phi_2 s} \left(X_t + \int_t^{t+s} \phi_1 e^{\phi_2 (u-t)} \, du + \int_t^{t+s} \xi e^{\phi_2 (u-t)} \, dW_u \right) ,$$

which leads to

$$\mathcal{L}(X_{t+s} - X_t | \mathcal{F}_t) = N\left(\mu(s, X_t), \sigma^2(s)\right) \tag{2.2}$$

where the conditional mean of the increment is

$$\mu(s, X_t) = \left(\frac{\phi_1}{\phi_2} - X_t \right) \left(1 - e^{-\phi_2 s} \right)$$

and the conditional variance[3]

$$\sigma^2(s) = \frac{\xi^2 \left(1 - e^{-2\phi_2 s}\right)}{2\phi_2} .$$

The asset market consists of a stock paying the dividend K and a market for borrowing and lending at a short rate (r_t).

The I agents have homogeneous expectations $P^i = P$ and constant relative risk aversion δ. They possibly differ in their time preferences f^i which are assumed to display a preferred habitat. Their utility function is thus

$$U^i(c) = E \int_0^T f^i(t) \frac{c_t^{1-\delta}}{1-\delta} \, dt .$$

For purposes of illustration, I take the case of two agents with logistic time preference densities where agent 1 has a short habitat ($h = 0$) and agent 2 a long habitat ($h > 0$).

[3] Note that μ and σ have a different meaning in Chapter 1. Since the objects they denoted there play no role in this chapter, I feel free to assign a new meaning to these symbols.

2.3 Equilibrium

Time has come to study the effects of preferred habitats on equilibrium consumption plans, prices and interest rates. It is important, as I said at the beginning of this chapter, to study the economy with heterogeneous agents, where different types of time preferences are present. A measure of heterogeneity is the vector s of the shares of stock the agents own initially. For every distribution of shares s there is a corresponding Pareto-efficient equilibrium characterized by a weight $\lambda = \lambda_s$ of some representative agent. I will compare the equilibrium in the heterogeneous economy with the equilibria of the homogeneous economies, in which only one type of agents is present. In homogeneous economies, the representative agent is clearly the only type of agent present in the economy, resulting in a no-trade equilibrium. Contrarily, in heterogeneous economies, trade occurs and more complicated allocations arise.

The representative agent's utility and the corresponding Pareto optimal allocations are calculated next.

Theorem 2.3.1. *For $\lambda \gg 0$ the utility function of the representative agent is*

$$\bar{U}(c, \lambda) = E \int_0^T f^\lambda(t) \frac{c_t^{1-\delta}}{1-\delta} dt$$

where the time preference density of the representative agent is given by

$$f^\lambda(t) = \left(\sum_{j=1}^{I} (\lambda^j f^j(t))^{\frac{1}{\delta}} \right)^{\delta} .$$

The corresponding Pareto efficient allocation is

$$c^i(\lambda)_t = \frac{(\lambda^i f^i(t))^{\frac{1}{\delta}}}{\sum_j (\lambda^j f^j(t))^{\frac{1}{\delta}}} K_t . \tag{2.3}$$

Proof. By (1.19) the felicity function of the representative agent is

$$v(t,x,\lambda) = \max_{\sum y^i = x} \sum_i \lambda^i f^i(t) \frac{(y^i)^{1-\delta}}{1-\delta}.$$

For some Lagrange multiplier ν, the first order condition

$$\lambda^i f^i(t)(y^i)^{-\delta} = \nu$$

holds true, hence

$$y^i = \left(\frac{\lambda^i f^i(t)}{\nu}\right)^{\frac{1}{\delta}}. \tag{2.4}$$

As a consequence,

$$x = \sum y^i = \frac{1}{\nu^{\frac{1}{\delta}}} \sum_i \left(\lambda^i f^i(t)\right)^{\frac{1}{\delta}},$$

and

$$\nu = \left(\sum_i \left(\lambda^i f^i(t)\right)^{\frac{1}{\delta}}\right)^{\delta} x^{-\delta}$$

follows. Going back to (2.4), one solves for

$$y^i = \frac{(\lambda^i f^i(t))^{\frac{1}{\delta}}}{\sum_j (\lambda^j f^j(t))^{\frac{1}{\delta}}} x,$$

which yields (2.3). Plugging in the optimal values of y^i, the value function results as

$$\begin{aligned} v(t,x,\lambda) &= \sum_i \lambda^i f^i(t) \left(\frac{(\lambda^i f^i(t))^{\frac{1}{\delta}}\, x}{\sum_j (\lambda^j f^j(t))^{\frac{1}{\delta}}}\right)^{1-\delta} \frac{1}{1-\delta} \\ &= \frac{\sum_i (\lambda^i f^i(t))^{\frac{1}{\delta}}}{\left(\sum_j (\lambda^j f^j(t))^{\frac{1}{\delta}}\right)^{1-\delta}} \frac{x^{1-\delta}}{1-\delta} \\ &= \left\|(\lambda^i f^i(t))_i\right\|_{\frac{1}{\delta}} \frac{x^{1-\delta}}{1-\delta}. \end{aligned}$$

In the Pareto optimal allocation characterized by the vector λ, the share agent i consumes out of total endowment is

$$\eta_t^i := \frac{c^i(\lambda)_t}{K_t} = \left(\frac{\lambda^i f^i(t)}{f^\lambda(t)}\right)^{\frac{1}{\delta}} .$$

The share η^i is deterministic and does not depend on the specific model for K. The share does depend on the relative weight

$$\frac{\lambda^i f^i(t)}{f^\lambda(t)}$$

agent i places on the point in time t with respect to the "market's" weight $f^\lambda(t)$. This shows that at a habitat, an agent will in general consume a greater share of total endowment than at other times. However, the share increases only if the market's time preference density does not rise at the same time, that is, if there are other people who do not have a preferred habitat at this point.

The time preference density of the representative agent is the length of the weighted vector of time preference densities $(\lambda^i f^i(t))_i$, where length is measured by the $\frac{1}{\delta}$-norm in $\mathbb{R}^I$. In the case of log-utility, $\delta = 1$, f^λ is just the weighted average $\sum \lambda^i f^i(t)$ of the time preference densities.

As usual, the rate of time preference of the representative agent is

$$\rho(\lambda)_t = -\frac{\partial}{\partial t} \log f^\lambda(t) = -\frac{\frac{\partial}{\partial t} f^\lambda(t)}{f^\lambda(t)} .$$

In the same manner,

$$\rho_t^i = -\frac{\partial}{\partial t} \log f^i(t)$$

is the rate of time preference of agent i.

Lemma 2.3.1. *The rate of time preference of the representative agent is a convex combination of the agents' time preferences:*

$$\rho(\lambda)_t = \sum_i \eta_t^i \rho_t^i ,$$

where the weights η^i are the shares of consumption of the agents.

Proof. This is an exercise in taking derivatives.

$$\begin{aligned}\frac{\partial}{\partial t}\log f^\lambda(t) &= \frac{\partial}{\partial t}\delta\log\left(\sum_{j=1}^{I}(\lambda^j f^j(t))^{\frac{1}{\delta}}\right)\\ &= \frac{\delta\sum_j(\lambda^j)^{\frac{1}{\delta}}\frac{1}{\delta}(\frac{\partial}{\partial t}f^j(t))f^j(t)^{\frac{1}{\delta}-1}}{\sum_{j=1}^{I}(\lambda^j f^j(t))^{\frac{1}{\delta}}}\\ &= \sum_i \frac{(\lambda^i f^i(t))^{\frac{1}{\delta}}}{\sum_{j=1}^{I}(\lambda^j f^j(t))^{\frac{1}{\delta}}}\frac{\frac{\partial}{\partial t}f^i(t)}{f^i(t)}\\ &= -\sum \eta^i(t)\rho^i(t)\,.\end{aligned}$$

As shown in the general theory of Chapter 1, the spot price of the consumption good is the marginal felicity of the representative agent, the short rate is minus the expected rate of growth of the spot price (cf. (1.27)), and the stock price S is given by the expected present value of the remaining dividends (cf. (1.29)).

Theorem 2.3.2. *An Arrow-Radner equilibrium with complete markets exists and is characterized by a vector $\lambda = \lambda(s)$ that is the solution of*

$$E\int_0^T \left(c^i(\lambda)_t - s^i K_t\right)\psi(\lambda)_t\,dt = 0 \quad i = 1,\dots,I \qquad (2.5)$$

The spot-price process for the consumption good is

$$\psi(\lambda)_t = f^\lambda(t)K_t^{-\delta}\,. \qquad (2.6)$$

The equilibrium short rate is

$$r(\lambda)_t = \rho(\lambda)_t + \delta(\phi_1 - \phi_2 X_t) - \frac{1}{2}\delta(1+\delta)\xi^2\,. \qquad (2.7)$$

The market price of risk p is the product of relative risk aversion and the volatility of X:

$$p = \delta\xi\,. \tag{2.8}$$

The stock price is

$$S(\lambda)_t = K_t \frac{\int_t^T \bar{g}(u,t,X_t)\,du}{f^\lambda(t)}, \tag{2.9}$$

where the risk-adjusted time preference density of the market $\bar{g}$ *is given by*

$$\bar{g}(u,t,x) = f^\lambda(u) e^{\left((1-\delta)\left(\mu(u-t,x)-\frac{\xi^2}{2}(u-t)\right)+\frac{(1-\delta)^2}{2}\sigma^2(u-t)\right)}.$$

The dynamics of the deflated gain process $G = G(\lambda)$ associated with $S(\lambda)$ are

$$dG(\lambda)_t = (1-\delta)\xi K_t^{1-\delta} \int_t^T e^{-\phi_2(u-t)} \bar{g}(u,t,X_t)\,du\,dW_t\,.$$

Proof. The general theory of Chapter 1 shows that an equilibrium is described by a vector λ which solves (2.5).

(2.6) is clear by (1.26) and Lemma 2.3.1.

To find the short rate and the market price of risk, apply Itô's lemma to ψ to obtain

$$\begin{aligned} d\psi_t &= d(f^\lambda(t)K_t^{-\delta}) \\ &= K_t^{-\delta} df^\lambda(t) + f^\lambda(t) dK_t^{-\delta} \\ &= -f^\lambda(t)K_t^{-\delta}\left(\left(-\frac{\frac{\partial}{\partial t}f^\lambda(t)}{f^\lambda(t)} + \delta(\phi_1 - \phi_2 X_t)\right.\right. \\ &\quad \left.\left. - \frac{\delta(1+\delta)\xi^2}{2}\right) dt + \delta\xi dW_t\right), \end{aligned}$$

hence

$$\frac{d\psi_t}{\psi_t} = -\left(\rho(\lambda)_t + \delta(\phi_1 - \phi_2 X_t) - \frac{1}{2}\delta(1+\delta)\xi^2\right) dt - \delta\xi dW_t$$

and (2.7) and (2.8) follow.

For the stock price, one uses the fact that the increment of the process X is normally distributed given the information up to time t, cf. (2.1) and (2.2). By (1.29)

$$\begin{aligned}
S_t &= \frac{1}{\psi_t} E\left[\int_t^T K_u \psi_u \, du \,\middle|\, \mathcal{F}_t\right] \\
&= \frac{K_t^\delta}{f^\lambda(t)} E\left[\int_t^T f^\lambda(u) K_u^{1-\delta} \, du \,\middle|\, \mathcal{F}_t\right] \\
&= \frac{K_t}{f^\lambda(t)} \int_t^T f^\lambda(u) E\left[\left(\frac{K_u}{K_t}\right)^{1-\delta} \,\middle|\, \mathcal{F}_t\right] du \\
&= \frac{K_t}{f^\lambda(t)} \int_t^T f^\lambda(u) E\left[e^{(1-\delta)(X_u - X_t) - \frac{1-\delta}{2}\xi^2(u-t))} \,\middle|\, \mathcal{F}_t\right] du,
\end{aligned}$$

and it remains to calculate the Laplace transform of a normal random variable:

$$E\left[e^{(1-\delta)(X_u - X_t)} \,\middle|\, \mathcal{F}_t\right] = e^{(1-\delta)\mu(u-t,t,X_t) + \frac{(1-\delta)^2}{2}\sigma^2(u-t)}.$$

In order to establish the completeness of the market, one has to check that Condition (1.31) holds true. Therefore, let us determine the diffusion coefficient $-\gamma$ of the gain process G. Note that by the preceding argument, the deflated stock price can be written as

$$S_t \psi_t = K_t^{1-\delta} D(t, X_t)$$

with

$$D(t,x) = \int_t^T \bar{g}(u,t,x)\, du.$$

Because of

$$\begin{aligned}
G_t &= S_t \psi_t + \int_0^t K_u \psi_u du \\
&= K_t^{1-\delta} D(t, X_t) + \int_0^t K_u \psi_u du,
\end{aligned}$$

the diffusion coefficient of G is the same as the diffusion coefficient of $K^{1-\delta}D$. The diffusion coefficient of $K_t^{1-\delta}$ is $(1-\delta)\xi K_t^{1-\delta}$. The diffusion coefficient of $D(t,X_t)$ is $\xi \partial/\partial x\, D(t,X_t)$. Using (2.2), one obtains

$$\frac{\partial}{\partial x} D(t,x) = \int_t^T (1-\delta) e^{-\phi_2(u-t)} \bar{g}(u,t,x) du - D(t,x).$$

The diffusion coefficient of the product $K^{1-\delta}D$ is therefore

$$\begin{aligned}
-\gamma_t &= (1-\delta)\xi D(t,X_t)K_t^{1-\delta} \\
&\quad + (1-\delta)\xi K_t^{1-\delta}\left(\int_t^T e^{-\phi_2(u-t)}\bar{g}(u,t,x)du - D(t,x)\right) \\
&= (1-\delta)\xi K_t^{1-\delta}\int_t^T (1-\delta)e^{-\phi_2(u-t)}\bar{g}(u,t,x)du \,. \quad (2.10)
\end{aligned}$$

Condition (1.31),

$$pS\psi \neq \gamma G$$

is therefore equivalent to

$$\delta \neq -(1-\delta)\frac{\int_t^T e^{-\phi_2(u-t)}\bar{g}(u,t,X_t)du}{\int_t^T \bar{g}(u,t,X_t)du} \,.$$

For $\delta \leq 1$ the right-hand side is nonpositive and the condition holds true. If $\delta > 1$, then the right-hand side is strictly smaller than δ since the fraction of integrals is smaller than 1 because of $\phi_2 \geq 0$ and $\delta - 1 < \delta$. Hence, the asset market is complete.

The short rate r is the sum of the rate of time preference $\rho(\lambda)$ and a component which does not depend on time preferences, but only on the risk aversion and the parameters of X,

$$\delta(\phi_1 - \phi_2 X_t) - \frac{1}{2}\delta(1+\delta)\xi^2 \,.$$

If one denotes by

$$r_t^i = \rho_t^i + \delta(\phi_1 - \phi_2 X_t) - \frac{1}{2}\delta(1+\delta)\xi^2$$

the short rate that prevails in the homogeneous economy where only agent i lives, then it follows as an obvious corollary of Lemma 2.3.1

Corollary 2.3.1 (Conjecture of Dumas). *The short rate of the heterogeneous world is a convex combination of the short rates that prevail in the homogeneous worlds,*

$$r(\lambda)_t = \sum_i \eta_t^i r_t^i .$$

In particular,

$$\min_i r_t^i \le r(\lambda)_t \le \max_i r_t^i .$$

The preceding corollary was conjectured in a paper of Dumas [23] who studied the case of heterogeneous degrees of risk aversion among agents. In the case of heterogeneous time preferences, it turns out that the short rate of the heterogeneous world is in between the bounds given by the short rates of the homogeneous economies and the conjecture of Dumas holds true(which is not always the case, cf. Wang [62]).

Since the rate of output growth X is a Brownian motion for $\phi_2 = 0$ and an autoregressive Ornstein-Uhlenbeck process for $\phi_2 > 0$, so is the short rate if the rate of time preference $\rho(\lambda)_t = \rho$ is constant. This type of model has been studied by Merton [46] (for the Brownian case) and by Vasiček[61] (for $\phi_2 > 0$). In general, $\rho(\lambda)$ varies with time and the resulting short rate is as in the models proposed by Hull and White [29]. Heterogeneity of time preferences can be seen as a reason for considering such models with a time-varying drift.

The next theorem gives the portfolio strategies agents use in equilibrium.

Theorem 2.3.3. *In equilibrium, agent i holds*

$$\theta_t^{i1} = \frac{\int_t^T \left((1-\delta)e^{-\phi_2(u-t)} + \delta\right) \eta_u^i \bar{g}(u,t,X_t)\, du}{\int_t^T \left((1-\delta)e^{-\phi_2(u-t)} + \delta\right) \bar{g}(u,t,X_t)\, du} \tag{2.11}$$

shares of the stock.

In the case of log-utility, $\delta = 1$,

$$\theta_t^{i1} = \frac{\lambda^i \int_t^T f^i(u)\, du}{\sum_j \lambda^j \int_t^T f^j(u)\, du} . \tag{2.12}$$

and there is no trading on the money market,

$$\theta_t^{i2} = 0 .$$

Proof. The general method for finding the portfolio to a given consumption stream is outlined in Theorem 1.5.1. Let

$$V_t^i = E\left[\int_t^T c_u^i \psi_u \, du \,\middle|\, \mathcal{F}_t\right]$$

be the deflated present value of the demanded consumption stream. By the familiar calculation

$$\begin{aligned} V_t^i &= E\left[\int_t^T \eta_u^i f^\lambda(u) K_u^{1-\delta} \, du \,\middle|\, \mathcal{F}_t\right] \\ &= K_t^{1-\delta} \int_t^T \eta_u^i \bar{g}(u,t,X_t) \, du \,. \end{aligned}$$

To find the portfolio, one has to know the diffusion coefficient m of V^i. For this purpose, one uses the same argument as in the proof of Theorem 2.3.2, where the diffusion coefficient of the gain process G is calculated, cf. (2.10). The result is:

$$dV_t^i = (1-\delta)\xi K_t^{1-\delta}\left(\int_t^T \eta_u^i e^{-\phi_2(u-t)} \bar{g}(u,t,X_t) du\right) dW_t \,.$$

The system of linear equations (1.35), (1.36) takes now the form

$$\begin{aligned} &\theta^{i1}(1-\delta)\xi K_t^{1-\delta} \textstyle\int_t^T e^{-\phi_2(u-t)} \bar{g}(u,t,X_t) du - \theta^{i2}\delta\xi H = \\ &(1-\delta)\xi K_t^{1-\delta} \textstyle\int_t^T \eta_u^i e^{-\phi_2(u-t)} \bar{g}(u,t,X_t) du \\ &\theta^{i1} K_t^{1-\delta} \textstyle\int_t^T \bar{g}(u,t,X_t) du + \theta^{i2} H = \\ &K_t^{1-\delta} \textstyle\int_t^T \eta_u^i \bar{g}(u,t,X_t) du \end{aligned}$$

The solution is given by (2.11).

If $\delta = 1$, then the first equation reduces to $-\theta^{i2}\delta\xi H = 0$, implying $\theta^{i2} = 0$. The equalities $\bar{g}(u,t,x) = f^\lambda(u)$ and $\eta^i f^\lambda = \lambda^i f^i$ yield (2.12).

The number of shares held by agent i at time t is given by the cumulated consumption shares η_u^i over the remaining time interval $[t, T]$ weighted with a certain density

$$\left((1-\delta)e^{-\phi_2(u-t)}+\delta\right)\bar{g}(u,t,X_t)\,.$$

This reflects the adjustment of the original time preference densities f^i due to risk aversion and the dynamics of X. In the case of log-utility, no adjustment is made and the number of shares is simply given by the fraction of the quantiles of the time preference distributions of agent i and the representative agent. The more weight an agent puts on the future, compared to the market's weight, the more she invests in the stock. Individuals with a long habitat will, therefore, initially defer consumption by investing in the stock (and re-investing the dividends obtained) in order to increase the number of shares of the stock in their portfolio. This strategy allows them to finance their higher consumption at their habitat. I repeat that this occurs only if there are other people who do the converse (have a short habitat). In a homogeneous economy, agents are forced to hold the stock at all times.

2.4 Analysis of the Term Structure

The general theory of Chapter 1 shows that the price of a zero-coupon bond is given by the present expected value of one unit of the consumption good at maturity, that is (cf. (1.44))

$$B_t^{t+\tau}=E\left[\frac{\psi_{t+\tau}}{\psi_t}\,\middle|\,\mathcal{F}_t\right].$$

Given the distributional assumptions of this chapter, it follows

Theorem 2.4.1. *Equilibrium bond prices are*

$$B_t^{\lambda,t+\tau}=\frac{f^\lambda(t+\tau)}{f^\lambda(t)}e^{-\delta\mu(\tau,X_t)+\delta\frac{\xi^2}{2}\tau+\frac{\delta^2}{2}\sigma^2(\tau)}\,. \tag{2.13}$$

The yield curve is therefore given by

$$y_t^{\lambda,t+\tau}=\frac{1}{\tau}\int_t^{t+\tau}\rho(\lambda)_u\,du+\delta\frac{\mu(\tau,X_t)}{\tau}-\delta\frac{\xi^2}{2}-\frac{\delta^2\sigma^2(\tau)}{2\tau}\,.$$

Proof. Bond prices are

$$B_t^{\lambda,t+\tau} = \frac{f^\lambda(t+\tau)}{f^\lambda(t)} E\left[\left(\frac{K_{t+\tau}}{K_t}\right)^{-\delta} \middle| \mathcal{F}_t\right]$$
$$= \frac{f^\lambda(t+\tau)}{f^\lambda(t)} E\left[\exp\left(-\delta(X_{t+\tau} - X_t) + \delta\frac{\xi^2}{2}\tau\right) \middle| \mathcal{F}_t\right],$$

Formula (2.13) follows using the conditional distribution of the increment $(X_{t+\tau}-X_t)$. For the yields, note that by definition of the rate of time preference $\rho(\lambda)$,

$$-\log\frac{f^\lambda(t+\tau)}{f^\lambda(t)} = \int_t^{t+\tau} \rho(\lambda)_u\,du\,.$$

Like the short rate, the long-term yields consist of two summands - the first,

$$\frac{1}{\tau}\int_t^{t+\tau} \rho(\lambda)_u\,du\,,$$

is due to the time preferences of the agents, whereas the second,

$$\delta\frac{\mu(\tau, X_t)}{\tau} - \delta\frac{\xi^2}{2} - \frac{\delta^2\sigma^2(\tau)}{2\tau},$$

is caused by the dynamics of the growth rate X and the relative risk aversion δ.

2.4.1 Illustration: Logistic Time Preferences

Up to here, the results are general and do not depend on the specific choice of the time preference densities f^i. To illustrate the effect of preferred habitats, it is assumed from now on that there are two agents ($I = 2$) with logistic time preference densities. Agent 1 is impatient, that is his habitat is $h^1 = 0$ and agent 2 has a long habitat, $h^2 > 0$.

Let us first assume, that the rate of growth X is a Brownian motion, that is $\phi_2 = 0$. Then $\mu(\tau, x) = \phi_1\tau$ and $\sigma^2(\tau) = \xi^2\tau$. The yield curve is

$$y_t^{t+\tau} = \frac{1}{\tau}\int_t^{t+\tau} \rho(\lambda)_u\,du + \delta\phi_1 - \frac{\delta(1+\delta)\xi^2}{2}.$$

The part caused by the dynamics of X and δ is constant and only the rates of time preference determine the shape of the yield curve. Interest rates are deterministic.

Let

$$f^i(t) = \frac{\exp(-\gamma^i(t-h^i))}{(1+\exp(-\gamma^i(t-h^i)))^2} \qquad i = 1,2$$

be the time preference densities of the agents and $F^i, i = 1,2$ the corresponding distribution functions.

Lemma 2.4.1. *The rate of time preference of the logistic time preference densities is*

$$\rho^i(t) = \gamma^i(2F^i(t) - 1)\,. \tag{2.14}$$

Proof. Note that F^i is a solution to the ordinary differential equation

$$f^i(t) = \gamma^i F^i(t)(1 - F^i(t))\,. \tag{2.15}$$

The logarithmic derivative is therefore

$$\begin{aligned}\frac{\partial}{\partial t}\log f^i(t) &= \frac{\partial}{\partial t}\left(\log\gamma^i + \log F^i(t) + \log(1-F^i(t))\right)\\ &= \frac{f^i(t)}{F^i(t)} - \frac{f^i(t)}{1-F^i(t)}\end{aligned}$$

and by applying (2.15), one obtains

$$= \gamma^i(1-F^i(t)) - \gamma^i F^i(t) = \gamma^i(1-2F^i(t))\,.$$

In homogeneous economies, the short rate is $r_t^i = \rho^i(t) + const$. It is increasing towards the level $\gamma^i + const.$, where γ^i is the long-run rate of time preference of the time preference density as defined in Definition 2.1.1. As the short rate is rising, so is the yield curve, since it is the average over the future short rates. As long as only one type of agent is present, the yield curve is always increasing, regardless of the habitat of the agents.

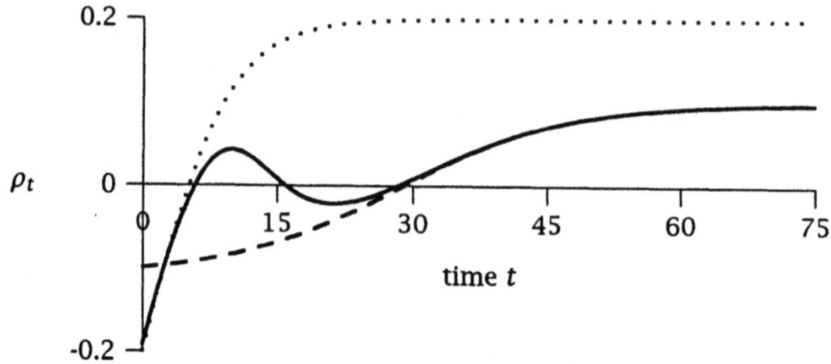

Fig. 2.2. Rates of time preference. Agents have log-utility. Agent 1 is impatient, $h^1 = 0$. The habitat of agent 2 is $h^2 = 30$ and the long-run rates of time preference are $\gamma^1 = 0.2$ and $\gamma^2 = 0.1$. ρ^1 is represented by a dotted line and ρ^2 by a dashed line. The rate of time preference $\rho(\frac{1}{2})$ of the representative agent with $\lambda = \frac{1}{2}$ is represented as a solid line.

The rate of time preference of an agent with a long habitat is the rate of an agent with a short habitat, shifted by the habitat h. Therefore, if the two long-run rates of time preference do not differ too much, the short rate in the homogeneous economy, where all agents display a short habitat, will be higher than short rate in the homogeneous economy, where all agents have a long habitat.

The shape of the yield curve changes only if different types of agents are present. By Lemma 2.3.1, in the heterogeneous economy, the rate of time preference $\rho(\lambda)$ is a time-varying average of the individuals' rates ,

$$\rho(\lambda)_t = \eta_t \rho_t^1 + (1 - \eta_t)\rho_t^2 ,$$

where $\eta = \eta^1$ is the share of consumption of agent 1 in equilibrium.

Two typical pictures of the mixed rate of time preference $\rho(\lambda)$ are shown in Figures 2.2 and 2.3. Initially, the impatient agent 1 consumes a large share of the aggregate endowment, as 0 is his habitat. Hence, η is close to 1, and the mixed rate $\rho(\lambda)$ is very close to the impatient agent's rate ρ^1. Later, the roles are reversed and agent 2 begins consuming a larger share of aggregate endowment. Around her habitat h_2, her share $1 -$

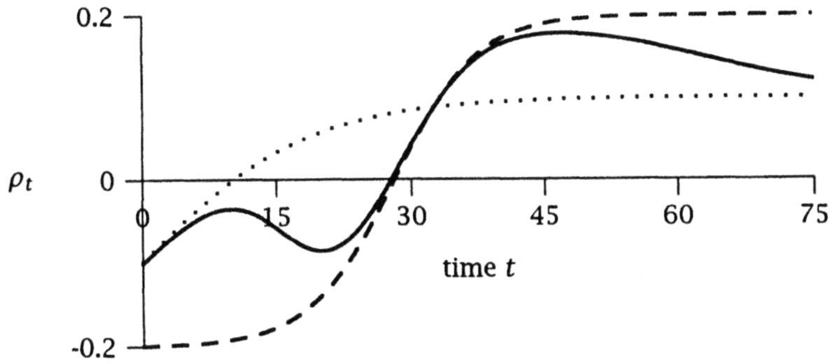

Fig. 2.3. Rates of time preference. Agents have log-utility. Agent 1 is impatient, $h^1 = 0$. The habitat of agent 2 is $h^2 = 30$ and the long-run rates of time preference are $\gamma^1 = 0.1$ and $\gamma^2 = 0.2$. ρ^1 is represented by a dotted line and ρ^2 by a dashed line. The rate of time preference $\rho(\frac{1}{2})$ of the representative agent with $\lambda = \frac{1}{2}$ is represented as a solid line.

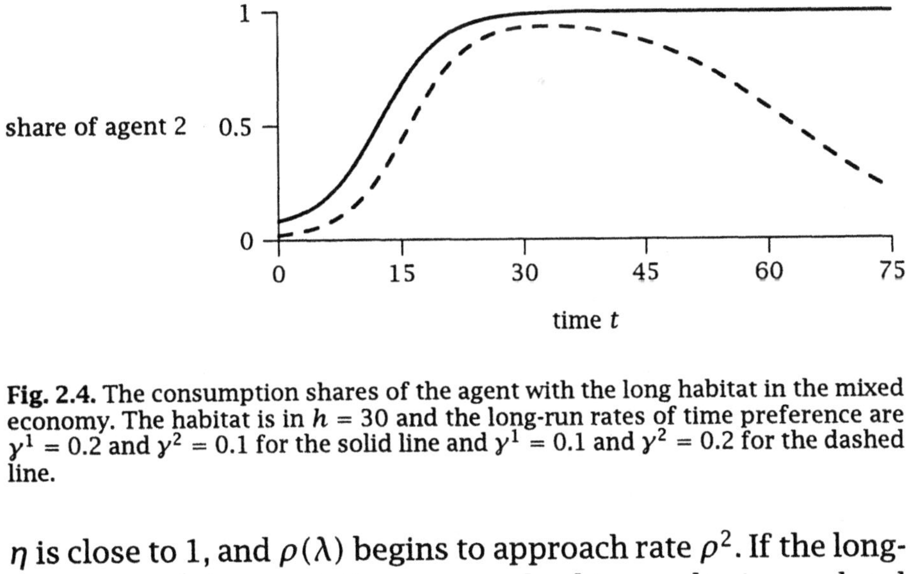

Fig. 2.4. The consumption shares of the agent with the long habitat in the mixed economy. The habitat is in $h = 30$ and the long-run rates of time preference are $\gamma^1 = 0.2$ and $\gamma^2 = 0.1$ for the solid line and $\gamma^1 = 0.1$ and $\gamma^2 = 0.2$ for the dashed line.

η is close to 1, and $\rho(\lambda)$ begins to approach rate ρ^2. If the long-run rates γ^i do not differ too much, this results in a a local maximum for $\rho(\lambda)$, since curve ρ^2 lies below curve ρ^1 .

In the long run, the behavior of $\rho(\lambda)$ depends on the long-run rates of time preference γ^i. The limit value of the share

$$\eta = \frac{(\lambda^1 f^1(t))^{\frac{1}{\delta}}}{(\lambda^1 f^1(t))^{\frac{1}{\delta}} + (\lambda^2 f^2(t))^{\frac{1}{\delta}}} = \left(1 + \left(\frac{\lambda^2 f^2(t)}{\lambda^1 f^1(t)}\right)^{\frac{1}{\delta}}\right)^{-1}$$

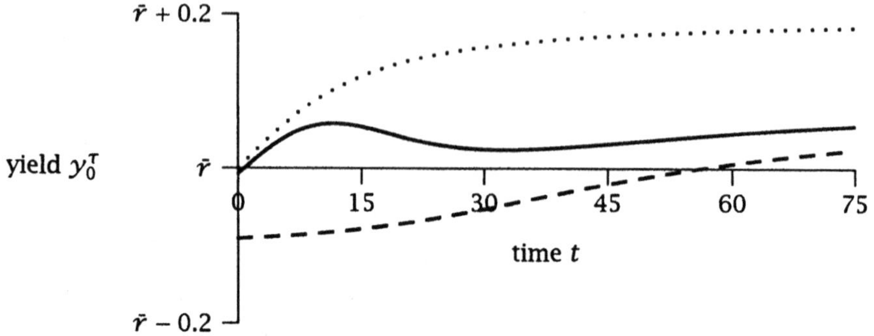

Fig. 2.5. The yield curves. The parameter values are as in Figure 2.2. In addition, $\phi_2 = 0$ is assumed. The yields fluctuate around the level $\bar{r} = \phi_1 - \xi^2$. The yield curves of the homogeneous economies, y^0 and y^1 are represented by dotted and dashed lines, respectively.

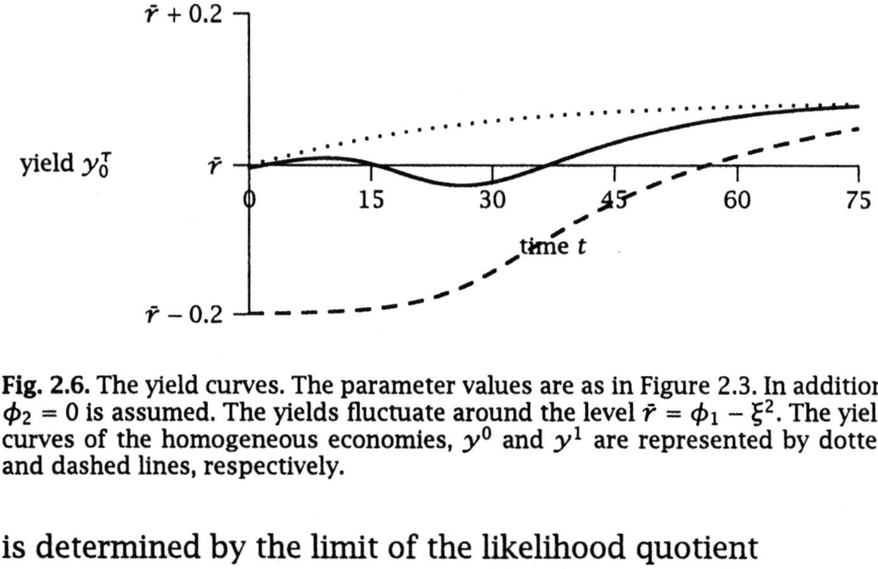

Fig. 2.6. The yield curves. The parameter values are as in Figure 2.3. In addition, $\phi_2 = 0$ is assumed. The yields fluctuate around the level $\bar{r} = \phi_1 - \xi^2$. The yield curves of the homogeneous economies, y^0 and y^1 are represented by dotted and dashed lines, respectively.

is determined by the limit of the likelihood quotient

$$\frac{f^2(t)}{f^1(t)} \sim \exp\left(-(y^2 - y^1)t\right) \quad (t \to \infty)\,.$$

Hence, η tends to 1 if $y^1 < y^2$ and to 0 if agent 2 is less impatient in the long run, $y^2 < y^1$. In the knife-edge case $y^1 = y^2$, both agents remain in the economy and $\eta_t \to \left(1 + \left(\frac{\lambda^2}{\lambda^1}\right)^{\frac{1}{\delta}}\right)^{-1}$. Typical curves of the share $1-\eta$ of agent 2 are given in Figure

2.4. The share starts at a relatively low level and increases up to the habitat. If agent 2 is less impatient, she also dominates in the long run. Her share continues to increase towards 1. Otherwise, her share decreases again to the long-run level of zero.

This has the following impact on the mixed rate of time preference: Near 0 the mixed rate $\rho(\lambda)$ follows the rate ρ^1 because agent 1 dominates, in the sense that she consumes a large part of the aggregate endowment. Later, around the habitat, agent 2 takes over. Therefore, the rate of time preference $\rho(\lambda)$ decreases to the lower rate ρ^1, attains a minimum, and then follows the curve ρ^1. If agent 2, with the long habitat, is less impatient in the long run, the mixed rate $\rho(\lambda)$ stays close to ρ^2 after the habitat. This case is depicted in Figure 2.2. If $\gamma^1 < \gamma^2$, $\rho(\lambda)$ again approaches ρ^1, see Figure 2.3.

The resulting yield curves, which are the average, up to a constant, of the rates of time preference, are shown in Figures 2.5 and 2.6. The averaged curve is naturally smoother than the original one. One sees a certain overshooting of the yields prior to the habitat and lower yields around the habitat - a hump appears in the yield curve. Therefore, in the heterogeneous economy, where people actually trade in equilibrium, preferred habitats for consumption create humps in the yield curve.

In order to isolate the effects of preferred habitats on the yield curve, the case of deterministic interest rates ($\phi_2 = 0$), where the yield curve would be flat neglecting time preferences, has been analyzed. In general, interest rates are stochastic. As already noted, apart the effect of time preferences, a Vasiček-type model is the outcome. It is well known that three types of yield curves, increasing, decreasing and single-humped, are possible in this case. The presence of heterogeneous preferred habitats causes an additive shift of the 'Vasiček'-curve and can lead to additional humps.

2.5 The Demand for Long-Term Bonds

Preferred habitats produce humps in the yield curve when there are different types of agents present in the economy. The resulting curves seem to indicate that there is a high demand for long-term bonds whose maturity correspond to a specific habitat. One may argue, though, that agents do not trade in bonds, since the market spanned by the stock and the money market account is complete. The question is whether they would buy long-term bonds if needed. For this reason, a variation of the model is studied, in which the risky stock is replaced with a long-term bond. Hence, the agents use the bond market to finance their (excess) demand, instead of the stock market, as in the preceding sections. The aim is to show that investors with a long-term habitat buy long-term bonds whereas their impatient counterparts do the converse.

The variation of the model goes as follows. There is no longer trade in the stock. Instead, the agents consider their endowment $e_t^i = s^i K_t$ as an exogeneously given income stream. According to their preferences, they form an excess demand $\epsilon_t^i = c_\lambda^i(\lambda)_t - e_t^i$ as in the Arrow-Debreu framework. To finance this excess demand, they use the bond market. More specifically, they trade in β, the money market account and B^T, the long-term bond whose maturity is the horizon T of the economy. As long as interest rates are stochastic, these two bonds will suffice to span a complete market. Therefore, $\phi_2 > 0$ is assumed, which yields stochastic interest rates. Moreover, log-utility is imposed, which simplifies matters considerably. The time preference densities are normalized to be probability densities on $[0, T]$ without loss of generality.

The first nice consequence of log-utility is that the equilibrium vector of weights $\lambda(s)$ is equal to the vector of initial shares s.

Lemma 2.5.1. *In the case of log-utility, the equilibrium allocation is described by the vector of weights*

$$\lambda^i = s^i .$$

Proof. In the case of log-utility,

$$f^\lambda(t) = \sum \lambda^i f^i(t),$$

$$\eta_t^i = \frac{\lambda^i f^i(t)}{f^\lambda(t)},$$

and

$$\psi(\lambda)_t = f^\lambda(t) K_t^{-1}.$$

In equilibrium, Equation (2.5) must hold. It simplifies here to

$$\begin{aligned} 0 &= E\int_0^T \left(c_\lambda^i(\lambda)_t - s^i K_t\right)\psi(\lambda)_t\, dt \\ &= E\int_0^T \left(\lambda^i f^i(t) - s^i f^\lambda(t)\right) dt \end{aligned}$$

implying $\lambda^i = s^i$ given that the time preference densities integrate to 1.

The next theorem is devoted to the dynamics of the long-term bond $B^T = B^{\lambda,T} = B^{s,T}$.

Theorem 2.5.1. *Bond price dynamics are*

$$\frac{dB_t^T}{B_t^T} = \left(r_t + \xi\sigma_t^T\right) dt + \sigma_t^T dW_t,$$

where the diffusion coefficient is

$$\sigma_t^T = \xi\left(1 - e^{-\phi_2(T-t)}\right).$$

Proof. One has to determine μ_t^T and σ_t^T such that $dB_t^T = B_t^T\left(\mu_t^T dt + \sigma_t^T dt\right)$. The deflated bond price process $M_t = B_t^T \psi_t = E\left[\psi_T | \mathcal{F}_t\right]$ is a martingale. By the product rule,

$$dM_t = M_t\left(\mu_t^T - r_t - p\sigma_t^T\right) dt + (\sigma_t^T - p) dW_t,$$

and

$$\mu_t^T = r + p\sigma_t^T$$

follows. Finally, $p = \xi$, since in the case of log-utility the coefficient of relative risk aversion is $\delta = 1$.

It remains to determine the diffusion coefficient σ_t^T. In light of (2.13), namely

$$B_t^T = \frac{f^\lambda(T)}{f^\lambda(t)} \exp\left(-(\mu(T-t, X_t) - \frac{\xi^2}{2}(T-t)) + \frac{1}{2}\sigma^2(T-t) \right),$$

it is clear that

$$\begin{aligned} \sigma_t^T &= -\frac{\partial}{\partial x}\mu(T-t, X_t)\,\xi \\ &= \left(1 - e^{-\phi_2(T-t)}\right)\xi . \end{aligned}$$

In order to calculate the portfolio strategies $(\theta^{1i}, \theta^{2i})$ in B^T and β that finance the excess demand ϵ^i, I proceed along the lines of Definition 1.2.1 and Theorem 1.5.1. One has just to replace the stock price S with the bond price B^T and the consumption demand c_λ^i with the excess demand ϵ^i.

Theorem 2.5.2. *To finance their excess demand ϵ^i, agents hold*

$$\theta_t^{1i} = \frac{s^i(F^s(t) - F^i(t))}{B_t^T \psi_t \left(1 - e^{-\phi_2(T-t)}\right)} \tag{2.16}$$

shares of the long-term bond.

In particular, an agent invests in the long-term bond if and only if she places more weight on the future than the market does:

$$1 - F^i(t) > 1 - F^s(t).$$

Proof. Let us determine the deflated value of the excess demand first:

$$\begin{aligned} V_t^i &:= E\left[\int_t^T \epsilon_u^i \psi_u \, du \,\middle|\, \mathcal{F}_t \right] \\ &= \int_t^T \left(s^i f^i(u) - s^i f^s(u)\right) du \\ &= s^i\left[(1 - F^i(t)) - (1 - F^s(i))\right] \\ &= s^i\left(F^s(t) - F^i(t)\right) . \end{aligned}$$

The martingale part of V^i is 0 because it is deterministic. Thus, a strategy that finances ϵ^i has to solve

$$\theta_t^{1i} B_t^T \psi_t + \theta_t^{2i} H_t = V_t^i$$
$$\theta_t^{1i} \left(\sigma_t^T - \xi\right) B_t^T \psi_t - \theta_t^{2i} \xi H_t = 0 .$$

The determinant of this system is

$$det = -B_t^T \psi_t H_t \left(1 - e^{-\phi_2 (T-t)}\right) < 0 .$$

It has therefore always a solution which is given by (2.16).

Now the question raised at the beginning of the section can be answered. If people are forced to trade in bonds, they really do buy long-term bonds if they have a long habitat. To have a long habitat means that the weight $1 - F^i(t)$ one's time preference distribution places on the future is greater than the average weight $\sum s^j \left(1 - F^j(t)\right)$. The intuition of Modigliani and Sutch that long term bonds are bought by agents with a long habitat is thus confirmed.

3. Imperfect Information: The Term Structure when the Growth Rate is Unknown

This chapter serves to illustrate the effect of imperfect information on the term structure of interest rates. To isolate the informational aspect, a homogeneous economy is studied in which all agents have the same type of preferences and beliefs. The output of the economy grows at a constant rate, which is known in the case of perfect information and unknown in the case of imperfect information. In the latter case, agents engage in Bayesian inference and choose their portfolio and consumption plan on the basis of the estimated growth rate. Since there is only one Brownian motion involved, the asset market turns out to be complete under perfect, as well as under imperfect information. It is therefore possible to compare the resulting equilibrium interest rates.

Given perfect information, the short rate is constant and the term structure remains flat at all times. The short rate, given imperfect information, is the conditional expectation of the short rate with perfect information and, therefore, a martingale. Whereas this is plausible, the resulting term structure is astonishing. It is always decreasing to the lowest possible level! In the specific case of normally distributed prior beliefs, the term structure is always linearly decreasing to negative infinity. In the general case of arbitrary priors, I show that the yield curve decreases to the essential infimum of the distribution of the short rate and that it is never above the actual short rate.

The effects of imperfect information on interest rates in a homogeneous world have already been studied in a number of papers by Detemple [14], Dothan and Feldman [18], and

Feldman [24]. All three model the growth rate as a stationary Ornstein-Uhlenbeck process. The resulting short rate with perfect information is also an Ornstein-Uhlenbeck process. In particular, interest rates are already stochastic with perfect information. The effect of imperfect information is that the diffusion coefficient of the short rate varies with time, whereas it was constant, given perfect information. In contrast to the approach of the cited authors, I start from a deterministic term structure and compare it with the case of stochastic interest rates. In the present model, stochastic interest rates are endogenously generated by the fact that the growth rate is unknown.

In the next section, I adapt the general model presented in Chapter 1 to the model sketched above. I describe the estimating procedure used by rational agents in Section 3.2 and in Section 3.3, the equilibrium stock price and short rate are derived. Finally, the last two sections are devoted to a study of the term structure. Section 3.4 gives explicit solutions for the yield curve in the case of normal or Bernoulli prior beliefs. The resulting yield curves are decreasing. This is not particular to the choice of specific prior beliefs, as I show in the last section. For arbitrary prior beliefs, it holds true that the yields converge to the lowest possible value in the long run, and that they are never above the short rate.

3.1 The Model

The growth rate X of aggregate output K is taken to be a Brownian motion with drift η,

$$\frac{dK_t}{K_t} = dX_t, \quad K_0 = 1$$
$$X_t = \eta t + W_t\,,$$

and W is a standard Brownian motion. Two cases are investigated. In the case of perfect information, agents know the value of the drift η. Their information filtration is

$$\mathcal{F}_t = \sigma\left(\eta, (X_s)_{s\leq t}\right) . \tag{3.1}$$

This case will be compared with the case of imperfect information where the agents see only the realizations of the growth rate X and use the information contained in

$$\mathcal{G}_t = \sigma\left((X_s)_{s\leq t}\right) \tag{3.2}$$

to estimate the drift η.

As I said above, all agents are identical in this chapter and share the same beliefs, time preferences and felicity functions. They all display constant relative risk aversion δ and discount uniformly the future with a constant rate of time preference $\rho > 0$. Thus, their utility function is

$$U^i(c) = E\int_0^T e^{-\rho t}\frac{c_t^{1-\delta}}{1-\delta}\,dt.$$

A particular role to play has agents' common prior belief $\mathcal{L}(\eta)$ about the distribution of the unknown drift. In the following section, it is assumed to be a normal or Bernoulli distribution. In the case of a normal prior, the yield curve decreases to negative infinity. Negative interest rates are ruled out by suitable assumptions on the support of the prior. I present the example of a Bernoulli prior, in which all yields are positive. Yet, the yield curve, given imperfect information, is also decreasing. A result for a general prior distribution is presented in Section 3.5.

3.2 Estimating the Drift

If the agents do not know the value of the drift η, they engage in Bayesian inference and obtain the estimator

$$\hat{\eta}_t = E\left[\eta \mid \mathcal{G}_t\right]$$

at time t. The cumulated estimated drift is $\int_0^t \hat{\eta}_u du$ and the observable remainder, which is not explained by the cumulated estimates, is the innovation process

$$\hat{W}_t = X_t - \int_0^t \hat{\eta}_u du .$$

By Lemma 1.3.1, $\hat{W}$ is an observable Brownian motion.

If the prior belief is a normal or Bernoulli distribution, $\hat{\eta}$ can be computed in closed form.

Lemma 3.2.1. *If agents have a normal prior belief,*

$$\mathcal{L}(\eta) \sim N(\mu, 1) ,$$

the a posteriori belief is

$$\mathcal{L}(\eta | X_t = x) = N\left(\frac{\mu + x}{1+t}, \frac{1}{1+t}\right) . \tag{3.3}$$

In particular,

$$\hat{\eta}_t = \frac{\mu + X_t}{1+t} . \tag{3.4}$$

The dynamics of $\hat{\eta}$ are given by

$$\hat{\eta}_t = \mu + \int_0^t \frac{1}{1+u} d\hat{W}_u . \tag{3.5}$$

If the prior belief is a Bernoulli distribution, that is

$$P(\eta = \mu_1) = p = 1 - P(\eta = \mu_0)$$

for some $\mu_1 > \mu_0$, the a posteriori belief $\hat{p}_t$ is

$$\hat{p}_t := P(\eta = \mu_1 | X_t = x) \tag{3.6}$$

$$= \left(1 + \frac{1-p}{p} \exp\left(-(\mu_1 - \mu_0)x + \frac{\mu_1^2 - \mu_0^2}{2} t\right)\right)^{-1} . \tag{3.7}$$

The estimate of the drift is thus

$$\hat{\eta}_t = (1 - \hat{p}_t)\mu_0 + \hat{p}_t \mu_1 ,$$

and its dynamics are given by

$$\hat{\eta}_t = (1-p)\mu_0 + p\mu^1 - \int_0^t v_t d\hat{W}_t ,$$

where the volatility v *is*

$$v_t = \frac{\frac{1-p}{p}(\mu_1-\mu_0)^2 e^{-(\mu_1-\mu_0)X_t + \frac{\mu_1^2-\mu_0^2}{2}t}}{\left(1+\frac{1-p}{p}e^{-(\mu_1-\mu_0)X_t+\frac{\mu_1^2-\mu_0^2}{2}t}\right)^2} . \qquad (3.8)$$

Proof. The current value X_t of the growth rate is a sufficient statistic for estimating the unknown drift η which yields $\hat{\eta}_t = E[\eta|X_t]$. Calculating $\hat{\eta}_t$ is then a standard exercise in Bayesian inference. The calculations are given for the normal case, the Bernoulli case is similar. Note that the conditional density of η given X_t is by Bayes' rule

$$\begin{aligned} P(\eta \in dz|X_t = x) &= \frac{P(X_t \in dx|\eta = z)P(\eta \in dz)}{P(X_t \in dx)} \\ &= \frac{\frac{1}{2\pi\sqrt{t}}\exp\left(-\frac{(x-zt)^2}{2t} - \frac{(z-\mu)^2}{2}\right)}{P(X_t \in dx)} . \qquad (3.9) \end{aligned}$$

Being the sum of two independent normal random variables, X is normally distributed. Thus, the total probability in the denominator is

$$P(X_t \in dx) = \frac{1}{\sqrt{2\pi t(1+t)}} \exp\left(-\frac{(x-\mu t)^2}{2t(1+t)}\right) .$$

Plugging this value into (3.9), one obtains the conditional distribution of η given $X_t = x$ as

$$\mathcal{L}(\eta|X_t = x) = N\left(\frac{\mu+x}{1+t}, \frac{1}{1+t}\right)$$

and $\hat{\eta}_t = \frac{\mu+X_t}{1+t}$ follows.

Apply Itô's formula (and neglect the drift term because $\hat{\eta}$ is a martingale!) to obtain the dynamics

$$d\hat{\eta}_t = \frac{1}{1+t} d\hat{W}_t .$$

3.3 Equilibrium with Perfect and Imperfect Information

In homogeneous economies, there is a no-trade equilibrium where the spot price of the consumption good is given by the marginal felicity of the one and only agent,

$$\psi_t = e^{-\rho t} K_t^{-\delta} .$$

Note that the dividend process K is observable with perfect as well as with imperfect information. Thus, the spot price for the consumption good is the same in both worlds. A straightforward calculation, which is already familiar from Chapter 2, yields the equilibrium asset prices with perfect information.

Theorem 3.3.1 (Perfect Information). *If the drift η is known, the equilibrium stock price is*

$$S_t = K_t \frac{1 - e^{-\zeta(T-t)}}{\zeta} , \tag{3.10}$$

with $\zeta = \rho + (1-\delta)(\frac{\delta}{2} - \eta)$[1]*. The short rate is*

$$r_t = \rho + \delta\eta - \frac{\delta(1+\delta)}{2} . \tag{3.11}$$

The market price of risk is equal to the degree of relative risk aversion,

$$p = \delta .$$

The asset market is complete.

Proof. The stock price is the deflated present value of the cumulated dividends:

$$\begin{aligned} S_t &= \frac{1}{\psi_t} E\left[\int_t^T K_u \psi_u \, du \,\middle|\, \mathcal{F}_t \right] \\ &= K_t E\left[\int_t^T e^{-\rho(u-t)} \left(\frac{K_u}{K_t} \right)^{1-\delta} du \,\middle|\, \mathcal{F}_t \right] \\ &= K_t \int_t^T e^{-\rho(u-t)} E\left[e^{(1-\delta)(X_u - X_t) - \frac{1-\delta}{2}(u-t)} \,\middle|\, \mathcal{F}_t \right] du . \end{aligned}$$

[1] If $\zeta = 0$ then $S_t = K_t(T-t)$ which is the limit of (3.10) as $\zeta \to 0$.

Given perfect information, η is known. Thus, the increment of X, $X_u - X_t = W_u - W_t + \eta(u-t)$, is independent of the past $\mathcal{F}_t$, and normally distributed with mean $\eta(u-t)$ and variance $u-t$. It follows

$$\begin{aligned} S_t &= K_t \int_t^T e^{-\rho(u-t) - \frac{1-\delta}{2}(u-t) + (1-\delta)\eta(u-t)} E\left[e^{(1-\delta)(W_u - W_t)}\right] du \\ &= K_t \int_0^{T-t} e^{-u\left(\rho + \frac{1-\delta}{2} - (1-\delta)\eta - \frac{(1-\delta)^2}{2}\right)} du \\ &= K_t \frac{1 - e^{-\zeta(T-t)}}{\zeta} \end{aligned}$$

with $\zeta = \rho + (1-\delta)(\frac{\delta}{2} - \eta)$.

The short rate and the market price of risk are determined via the dynamics of the spot price process ψ (cf. (1.28)).

To check that the asset market spanned by the stock and the money market account is complete, one has to determine the diffusion coefficient of the gain process $G_t = S_t\psi_t + \int_0^t K_u\psi_u du$. As in the proof of Theorem 2.3.2, it suffices to calculate the diffusion coefficient of $S_t\psi_t = e^{-\rho t}\frac{1-e^{-\zeta(T-t)}}{\zeta}K_t^{1-\delta}$ which is $S_t\psi_t(1-\delta)$. The completeness condition (1.31) simplifies here to $\delta - 1 \neq \delta$ which holds always true.

With imperfect information, agents use the innovation process $\hat{W}$ instead of the underlying, but unobservable Brownian motion W in their calculations. In particular,

$$\frac{dK_t}{K_t} = dX_t = \hat{\eta}_t dt + d\hat{W}_t \,. \tag{3.12}$$

The same argument as in the preceding proof yields the equilibrium relations in the world with unknown drift η.

Theorem 3.3.2 (Imperfect Information). *If the drift η is unknown, the equilibrium stock price is the estimator of the equilibrium stock price with perfect information:*

$$\hat{S}_t = E\left[S_t \mid \mathcal{G}_t\right] = K_t E\left[\frac{1 - e^{-\zeta(T-t)}}{\zeta} \,\middle|\, \mathcal{G}_t\right]. \tag{3.13}$$

The market price of risk is equal to the degree of relative risk aversion, $p = \delta$.

The short rate is the estimator of the short rate with perfect information, thus a martingale:

$$\hat{r}_t = E\left[r \mid \mathcal{G}_t\right] = \rho + \delta\hat{\eta}_t - \frac{\delta(1+\delta)}{2}. \tag{3.14}$$

Proof. The short rate and the market price of risk are determined by applying Itô's formula to the consumption price process ψ. With the aid of (3.12),

$$-\frac{d\psi_t}{\psi_t} = \left(\rho + \delta\hat{\eta}_t - \frac{\delta(1+\delta)}{2}\right) dt + \delta d\hat{W}_t .$$

The candidate for the equilibrium stock price is

$$\begin{aligned}
\hat{S}_t &= E\left[\frac{1}{\psi_t}\int_t^T K_u\psi_u\, du \,\middle|\, \mathcal{G}_t\right] \\
&= E\left[E\left[\frac{1}{\psi_t}\int_t^T K_u\psi_u\, du \,\middle|\, \mathcal{F}_t\right] \middle|\, \mathcal{G}_t\right] \\
&= E\left[S_t \mid \mathcal{G}_t\right] .
\end{aligned}$$

With specific assumptions about the prior belief, the dynamics of the short rate can be computed explicitly:

Corollary 3.3.1. *If the prior belief about η is normal, the dynamics of the equilibrium short rate $\hat{r}$ are*

$$d\hat{r}_t = \delta d\hat{\eta}_t = \frac{\delta}{1+t} d\hat{W}_t .$$

In the Bernoulli case,

$$d\hat{r}_t = \delta v_t d\hat{W}_t ,$$

with v given by (3.8).

Proof. Immediately from the preceding theorem and Lemma 3.2.1.

For a fixed horizon T, the short rate $\hat{r}_t$ is determined for all times $t \leq T$. Since the short rate is independent of the horizon T, it is possible to obtain a process $(\hat{r}_t)_{0 \leq t < \infty}$ for all times t by letting the horizon T go to infinity. The process thus obtained converges to the 'true' but unknown value r:

Theorem 3.3.3.

$$\lim_{t \to \infty} \hat{r}_t = r \quad a.s.$$

Proof. It is sufficient to show that the Bayesian estimator $\hat{\eta}$ is consistent, that is

$$\lim \hat{\eta}_t = \eta \quad a.s.$$

By a martingale convergence theorem of Lévy, one has

$$\lim \hat{\eta}_t = E\left[\eta | \mathcal{G}_\infty\right] \quad a.s.$$

The estimator $\frac{X_t}{t}$ is a consistent estimator of η because of the strong law of large numbers for Brownian motion :

$$\lim \frac{X_t}{t} = \eta \quad a.s.$$

Hence, it suffices to show,

$$\hat{\eta}_t - \frac{X_t}{t} \to 0$$

in probability. Note that

$$\begin{aligned} \hat{\eta}_t - \frac{X_t}{t} &= E\left[\eta - \frac{X_t}{t} \middle| \mathcal{G}_t\right] \\ &= E\left[\frac{W_t}{t} \middle| \mathcal{G}_t\right], \end{aligned}$$

hence

$$E\left(\hat{\eta}_t - \frac{X_t}{t}\right)^2 \leq E\left(\frac{W_t}{t}\right)^2 = \frac{1}{t} \to 0.$$

3.4 The Yield Curve with Normal and Bernoulli Prior Beliefs

Up to this point, the results seem to be as one expects them. The unexpected effect appears if one looks at the term structure. Given perfect information, the short rate r is known at time 0 and remains constant for all times. Therefore, the term structure is flat,

$$y_t^{t+\tau} = r = \rho + \delta\eta - \frac{(1+\delta)\delta}{2}, \tag{3.15}$$

where $y_t^{t+\tau}$ denotes the yield to maturity $t+\tau$ at time t. It is the purpose of the remaining part of the chapter to show that, with imperfect information, the term structure is a decreasing function. As a first illustration, it is shown that, if the prior is normal, the yield curve is linear and decreases to negative infinity!

Theorem 3.4.1. *If the common prior belief about η is normal,*

$$\mathcal{L}(\eta) \sim N(\mu, 1),$$

the term structure is given by

$$\widehat{y_t^{t+\tau}} = \hat{r}_t - \frac{\delta^2}{2}\frac{\tau}{1+t}. \tag{3.16}$$

Proof. Let $\widehat{B_t^{t+\tau}}$ be the price of a zero-coupon bond with maturity $t+\tau$ at time t. By (1.44),

$$\widehat{B_t^{t+\tau}} = E\left[\frac{\psi_{t+\tau}}{\psi_t}\,\middle|\, \mathcal{G}_t\right].$$

Hence,

$$\begin{aligned}\widehat{B_t^{t+\tau}} &= E\left[\exp\left(-\rho\tau - \delta\left(X_{t+\tau} - X_t - \frac{1}{2}\tau\right)\right)\middle|\, \mathcal{G}_t\right]\\ &= E\left[e^{-\rho\tau - \delta\eta\tau + \frac{\delta}{2}\tau} E\left[\exp\left(-\delta(W_{t+\tau} - W_t)\right) \mid \mathcal{F}_t\right]\middle|\, \mathcal{G}_t\right]\\ &= E\left[\exp\left(-\rho\tau - \delta\eta\tau + \frac{\delta}{2}\tau + \frac{\delta^2}{2}\tau\right)\middle|\, \mathcal{G}_t\right].\end{aligned}$$

It remains to compute the conditional Laplace transform of the unknown drift η given $\mathcal{G}_t$. By Lemma 3.2.1, the conditional distribution of η given $\mathcal{G}_t$ is normal with mean $\hat{\eta}_t$ and variance $\frac{1}{1+t}$. Therefore,

$$\begin{aligned}\widehat{B_t^{t+\tau}} &= \exp\left(-\rho\tau - \delta\hat{\eta}_t\tau + \frac{\delta^2}{2}\frac{\tau^2}{1+t} + \frac{\delta}{2}\tau + \frac{\delta^2}{2}\tau\right)\\ &= \exp\left(-\tau(\rho + \delta\hat{\eta}_t - \frac{\delta^2}{2}\frac{\tau}{1+t} - \frac{\delta(1+\delta)}{2})\right).\end{aligned}$$

One obtains the yields

$$\begin{aligned}\widehat{y_t^{t+\tau}} &= -\frac{1}{\tau}\log\widehat{B_t^{t+\tau}}\\ &= \hat{r}_t - \frac{\delta^2}{2}\frac{\tau}{1+t}.\end{aligned}$$

Normal prior beliefs lead to negative interest rates in the long run, which implies that bond prices *increase* with maturity. Here, one even has $B_t^{t+\tau} \to_\tau \infty$. A dollar paid in the long distant future has an arbitrarily large present value. A very counterintuitive result.

It is possible to preclude negative interest rates and increasing bond prices by imposing a suitable lower bound on the prior belief. To give an example, I study the case of a Bernoulli prior belief,

$$P(\eta = \mu_1) = p = 1 - P(\eta = \mu_0) \tag{3.17}$$

with $\mu_1 > \mu_0$ and

$$y_{min} := \rho + \delta\mu_0 - \frac{\delta(1+\delta)}{2} \geq 0\,. \tag{3.18}$$

Then, as the next theorem shows, all yields are nonnegative. However, the term structure is still decreasing.

Theorem 3.4.2. *With a Bernoulli prior as in (3.17) and (3.18), all yields are nonnegative:*

$$\widehat{y_t^{t+\tau}} \geq y_{min} \geq 0 .$$

The term structure is explicitly given by

$$\widehat{y_t^{t+\tau}} = y_{min} - \frac{1}{\tau} \log \left(\hat{p}_t e^{-\delta(\mu_1 - \mu_0)\tau} + (1 - \hat{p}_t) \right) .$$

It is decreasing with long-run limit y_{min}.

Proof. As in the proof of Theorem 3.4.1, bond prices are given by the following conditional Laplace transform:

$$\widehat{B_t^{t+\tau}} = E\left[\exp\left(-\rho\tau - \delta\eta\tau + \frac{\delta}{2}\tau + \frac{\delta^2}{2}\tau \right) \middle| \mathcal{G}_t \right] .$$

The conditional distribution of η in the case of a Bernoulli prior has been computed in Lemma 3.2.1. Hence,

$$\begin{aligned} \widehat{B_t^{t+\tau}} &= \hat{p}_t \exp\left(-\rho\tau - \delta\mu_1\tau + \frac{\delta}{2}\tau + \frac{\delta^2}{2}\tau \right) \\ &\quad + (1 - \hat{p}_t) \exp\left(-\rho\tau - \delta\mu_0\tau + \frac{\delta}{2}\tau + \frac{\delta^2}{2}\tau \right) \\ &= e^{-\rho\tau - \delta\mu_0\tau + \frac{\delta}{2}\tau + \frac{\delta^2}{2}\tau} \left(\hat{p}_t e^{-\delta(\mu_1 - \mu_0)\tau} + (1 - \hat{p}_t) \right) \\ &= e^{-\tau y_{min}} \left(\hat{p}_t e^{-\delta(\mu_1 - \mu_0)\tau} + (1 - \hat{p}_t) \right) . \end{aligned}$$

Yields are therefore

$$\begin{aligned} \widehat{y_t^{t+\tau}} &= -\frac{1}{\tau} \log \widehat{B_t^{t+\tau}} \\ &= y_{min} - \frac{1}{\tau} \log \left(\hat{p}_t e^{-\delta(\mu^1 - \mu^0)\tau} + (1 - \hat{p}_t) \right) . \end{aligned}$$

Because of $\mu_1 > \mu_0$, the argument of the logarithm is smaller than 1, $\hat{p}_t e^{-\delta(\mu^1 - \mu^0)\tau} + (1 - \hat{p}_t) < 1$, hence all yields are nonnegative: $\widehat{y_t^{t+\tau}} \geq y_{min} \geq 0$. Moreover, $e^{-\delta(\mu^1 - \mu^0)\tau}$ tends to zero as maturity increases inducing

$$\lim_{\tau \to \infty} \widehat{y_t^{t+\tau}} = y_{min} .$$

To see that the yield curve is decreasing, one has to differentiate,

$$\frac{\partial}{\partial \tau}\widehat{y_t^{t+\tau}} = \frac{\log\left(\hat{p}_t e^{-\delta(\mu^1-\mu^0)\tau} + (1-\hat{p}_t)\right)}{\tau^2} + \frac{\hat{p}_t \delta(\mu_1 - \mu_0) e^{-\delta(\mu^1-\mu^0)\tau}}{\tau\left(\hat{p}_t e^{-\delta(\mu^1-\mu^0)\tau} + (1-\hat{p}_t)\right)}.$$

Using the inequality

$$\log(1-x) < -\frac{x}{1-x} \quad \text{for } 0 < x < 1,$$

one obtains

$$\begin{aligned}
\frac{\partial}{\partial \tau}\widehat{y_t^{t+\tau}} &< -\frac{\hat{p}_t\left(1 - e^{-\delta(\mu^1-\mu^0)\tau}\right)}{\tau^2\left(\hat{p}_t e^{-\delta(\mu^1-\mu^0)\tau} + (1-\hat{p}_t)\right)} \\
&\quad + \frac{\hat{p}_t \delta(\mu_1 - \mu_0) e^{-\delta(\mu^1-\mu^0)\tau}}{\tau\left(\hat{p}_t e^{-\delta(\mu^1-\mu^0)\tau} + (1-\hat{p}_t)\right)} \\
&= \frac{\hat{p}_t\left(\tau\delta(\mu_1-\mu_0)e^{-\delta(\mu^1-\mu^0)\tau} - \left(1 - e^{-\delta(\mu^1-\mu^0)\tau}\right)\right)}{\tau^2\left(\hat{p}_t e^{-\delta(\mu^1-\mu^0)\tau} + (1-\hat{p}_t)\right)} \\
&= \frac{\hat{p}_t\left(-1 + e^{-\delta(\mu^1-\mu^0)\tau}\left(1 + \tau\delta(\mu_1-\mu_0)\right)\right)}{\tau^2\left(\hat{p}_t e^{-\delta(\mu^1-\mu^0)\tau} + (1-\hat{p}_t)\right)}.
\end{aligned}$$

Because of

$$e^{-x} < \frac{1}{1+x} \quad \text{for positive } x,$$

the term in brackets is negative,

$$-1 + e^{-\delta(\mu^1-\mu^0)\tau}\left(1 + \tau\delta(\mu_1-\mu_0)\right) < 0,$$

and

$$\frac{\partial}{\partial \tau}\widehat{y_t^{t+\tau}} < 0$$

follows.

3.5 General Prior Beliefs

One might wonder whether the astonishing result of the preceding section is due to the choice of particular prior beliefs. This is not the case. In this section, I consider the case of a general prior belief whose essential infimum is finite. It is shown that with imperfect information, the yield curve is always bounded above by its initial value $\widehat{y_t^t} = \hat{r}_t$, the short rate. Moreover, it decreases to the lowest possible value for the interest rates.

Let $m = \operatorname{ess\,inf} \eta > -\infty$ be the finite essential infimum of the prior belief and assume $Var(\eta) > 0$. A representative agent equilibrium is given by Theorems 3.3.1 and 3.3.2. The equilibrium short rate with imperfect information is given by the expected value of the short rate with perfect information,

$$\hat{r}_t = E[r \mid \mathcal{G}_t] = \rho + \delta\hat{\eta}_t - \frac{\delta(1+\delta)}{2}.$$

Since $\hat{\eta}_t \geq \operatorname{ess\,inf} \eta = m$, the short rate and all yields are bounded below by $y_{min} = \rho + \delta m - \frac{\delta(1+\delta)}{2}$. The announced result is

Theorem 3.5.1. *Let $-\infty < m = \operatorname{ess\,inf} \eta$ and $Var(\eta) > 0$. Then the yield curve is bounded above by the short rate and bounded below by $y_{\min}$,*

$$y_{\min} \leq \widehat{y_t^{t+\tau}} \leq \hat{r}_t$$

and converges to the lower bound $y_{\min}$:

$$\lim_{\tau \to \infty} \widehat{y_\tau^{t+\tau}} = y_{\min} \,.$$

Proof. From the proofs of Theorems 3.4.1 and 3.4.2, it is known that bond prices are equal to a conditional Laplace transform:

$$\widehat{B_t^{t+\tau}} = e^{-\rho\tau + \frac{\delta}{2}\tau + \frac{\delta^2}{2}\tau} E[\exp(-\delta\eta\tau) \mid \mathcal{G}_t] \,.$$

The yields are therefore

$$\begin{aligned}\widehat{y_t^{t+\tau}} &= \rho - \frac{\delta}{2}(1+\delta) - \frac{1}{\tau}\log E\left[\exp(-\delta\eta\tau)\,|\,\mathcal{G}_t\right] \\ &\le \rho - \frac{\delta}{2}(1+\delta) + \delta\hat{\eta}_t = \hat{r}_t\end{aligned}$$

where Jensen's inequality has been used. The yield curve is bounded by its initial value.

For the long-run value of the yields, let the horizon T tend to infinity to obtain a whole family $\left(y_t^{t+\tau}\right)_{\tau\ge 0}$ of yields from 0 up to ∞. Denote by

$$L(\lambda) = E\left[\exp(-\lambda\eta)\,|\,\mathcal{G}_t\right]$$

the conditional Laplace transform of the unknown drift η. Lemma 3.5.1 below shows that the essential infimum of the conditional distribution of η is equal to the essential infimum of the distribution of η, namely m. The result, $\widehat{y_t^{t+\tau}} \to y_{min}$ follows then from the Lemma 3.5.2 on Laplace transforms.

The following lemmata complete the proof of Theorem 3.5.1:

Lemma 3.5.1. *The essential infimum of the conditional distribution of η given X_t is m,*

$$\operatorname{ess\,inf} \mathcal{L}\left(\eta|X_t\right) = \operatorname{ess\,inf} \eta = m\,.$$

Proof. The result is due to the fact that a Brownian motion can assume all real values. Therefore, on the basis of X_t, it is impossible to exclude a value for η with certainty if this value is initially possible.

By Bayes' rule,

$$P\left(\eta \le \alpha|X_t = x\right) = \frac{P\left(X_t \in dx|\eta \le \alpha\right)P\left(\eta \le \alpha\right)}{P\left(X_t \in dx\right)}\,.$$

It follows that

$$P\left(\eta \le \alpha|X_t = x\right) > 0$$

if and only if

$$P(\eta \leq \alpha) > 0.$$

This implies the result.

It remains to establish the lemma on Laplace transforms.

Lemma 3.5.2 (on Laplace transforms). *Let Y be a random variable and $m > -\infty$ its essential infimum. Then the Laplace transform*

$$L_Y(\lambda) = E\exp(-\lambda Y)$$

exists for all nonnegative λ. The asymptotic behavior of L_Y is given by

$$\lim_{\lambda\to\infty} -\frac{1}{\lambda}\log L_Y(\lambda) = m. \tag{3.19}$$

Proof. Without loss of generality let $m = 0$. Because of $Y \geq 0$ *a.e.*, one has $e^{-\lambda Y} \leq 1$ *a.e.* and by monotonicity of the integral

$$-\frac{1}{\lambda}\log L_Y(\lambda) \geq 0.$$

Assume that

$$\limsup -\frac{1}{\lambda}\log L_Y(\lambda) = \epsilon > 0.$$

Then there exists a sequence (λ_n) with $\lambda_n \to \infty$ and

$$-\frac{1}{\lambda_n}\log L_Y(\lambda_n) \geq \frac{\epsilon}{2} \quad \text{for all } n.$$

This is equivalent to

$$Ee^{-\lambda_n Y} \leq e^{-\lambda_n \frac{\epsilon}{2}}$$

or, multiplying both sides with $e^{\lambda_n \frac{\epsilon}{4}}$,

$$Ee^{-\lambda_n (Y-\frac{\epsilon}{4})} \leq e^{-\lambda_n \frac{\epsilon}{4}}.$$

Fatou's lemma gives

$$E \liminf e^{-\lambda_n (Y - \frac{\epsilon}{4})} \le \lim E e^{-\lambda_n (Y - \frac{\epsilon}{4})} \le \lim e^{-\lambda_n \frac{\epsilon}{4}} = 0 \, .$$

This implies

$$e^{-\lambda_{n_k} (Y - \frac{\epsilon}{4})} \to 0 \quad a.e.$$

for a subsequence (λ_{n_k}), which is only possible if

$$Y \ge \frac{\epsilon}{4} \quad a.e.,$$

a contradiction.

The preceding Theorem 3.5.1 shows that the puzzling result derived does not depend on specific assumptions about the prior belief. It is generally true in this model that the term structure is flat with perfect information, whereas the lack of information about the growth rate leads to a decreasing yield curve. One would expect the converse, though. With imperfect information, interest rates are stochastic, and long-term bonds are risky assets, in the sense that their price fluctuates randomly, even though their final value is fixed. The usual Hicksian argument already encountered in Chapter 2 states that risk averse investors require a positive risk premium for such assets. However, positive risk premia imply increasing yield curves whereas the converse is the case! Thus, the Hicksian argument is not valid here.

I offer the following explanation: within the model presented in this chapter, the only important uncertainty stems from the growth rate. If it is low, aggregate output may decrease very quickly. In this case, investment in a long term bond is a good hedge against unfavorable realizations of the growth rate. Due to such uncertainty, the demand for long term bonds increases, and yields are consequently lower.

4. Bulls and Bears: Heterogeneous Expectations

"It is differences in opinion that makes horse races", Mark Twain says. Following this saying, I study differences in expectations among agents in this chapter. Agents conform that the state of the world is described by a diffusion model as in Chapter 1. They differ, however, in their prior beliefs about some unknown parameters of the model. To give an example, the drift of the growth rate may be unknown, as in the preceding chapter, and some agents, call them bulls, might hold more optimistic views about this drift than others.

Due to the initial divergence of beliefs, agents obtain different estimates of the unknown parameters. As a consequence, agents trade with each other[1]. Bulls borrow in the money market in order to invest in the stock market, whereas bears sell shares of the stock to invest in the money market.

The consumption share of an agent is given by a likelihood fraction, as in Chapter 2 on heterogeneous time preferences. Here, it is the likelihood fraction of the individual's belief with respect to the market's belief which determines the equilibrium consumption share of an agent. It is interesting to note

[1] The reader familiar with the literature on no-trade theorems may feel uncomfortable at this point. Has it not been shown that rational agents never trade solely on the basis of differences in beliefs? The literature on the issue, Aumann [6], Kreps [37], Milgrom and Stokey [49], and Tirole [59], assumes a common prior for all agents. Heterogeneity of *a posteriori* beliefs is induced by asymmetric information. Then, in a rational expectations equilibrium, there is no trade. The reason for this result is, that agents would be willing to trade only, if they expected a positive monetary gain from trading. With a common prior, though, it is impossible to have positive expected gains for everyone, since the asset market is a zero sum game. In my model, however, information is symmetric among agents, but the priors differ. With heterogeneous priors, expected gains from trade for all agents become possible, and agents trade in equilibrium.

that the objective measure P has no influence on consumption beyond determining the equivalence class of possible beliefs, as the equilibrium relations depend only on the set of subjective beliefs (P^i).

As far as interest rates are concerned, one obtains several weighted average characterizations, similar to Chapter 2. The short rate is a weighted average of the short rates which prevail in homogeneous economies populated by one type of investor only. The weights are the likelihood fractions just mentioned which determine also the consumption shares. The same characterization holds true for bond prices. It follows that forward rates are also weighted averages of the forward rates of homogeneous economies. The weights, however, are different; they are given by the product of likelihood fractions and bond prices.

A wider variety of term structure shapes is possible in a world with heterogeneous beliefs. In two specific cases, the curves are explicitly calculated. In the first example, I take up the model studied in Chapter 3 with an unknown drift of the growth rate. Yield curves have a different shape with heterogeneous beliefs, but are still decreasing. In the second example, yield curves are computed for the case of an autoregressive specification of the unknown drift of the growth rate.

A similar study on asset pricing with heterogeneous beliefs and log-felicity has been carried out by Detemple and Murthy [15] in the context of a production economy, where the price of the physical asset is exogenously given. In contrast to my model, they have an incomplete asset market because they introduce an information technology which is not traded and can not be hedged. Their results for the short rate and asset prices are similar to mine. They also obtain a weighted average characterization of the short rate and of the bond prices. The weights they give are the fractions of the wealth of the agents. In my model I show that the fraction of wealth of an agent and total wealth of the economy is equal to the likelihood fraction of beliefs mentioned above. Hence, I offer a further interpretation of the weights. Another point of difference is the method used to calculate the equilibrium. I use the Negishi method to obtain the equilibrium, whereas Detemple and Murthy employ the stochastic control methods developed by Karatzas et al. [32] to calculate the demands which are then aggregated.

4.1 Setup

I describe now the setup for this chapter and recall some of the results derived in Chapter 1. The diffusion setup for aggregate consumption is kept completely general as in Chapter 1, cf. Section 1.2. For the reader's convenience, I recall that aggregate output K is given by

$$\frac{dK_t}{K_t} = dX_t, \quad K_0 = 1, \tag{4.1}$$

where the growth rate X evolves as

$$dX_t = \phi(X,Y)_t dt + \xi(X)_t dW_t, \quad X_0 = 0\,. \tag{4.2}$$

The factors Y are unknown parameters.

Agents' preferences are specified to be

$$U^i(c) = E^i \int_0^T f(t)\log(c_t)dt\,,$$

where the time preference density f is the same for all agents and a smooth probability density on $[0,T]$ with distribution function F. Let $\rho(t) = -\frac{\partial}{\partial t}\log f(t)$ be agents' rate of time preference. The agents differ in their beliefs about the structural parameter η,

$$\mathcal{L}\left(\eta|P^i\right) \neq \mathcal{L}\left(\eta|P^j\right), \quad (i \neq j)\,.$$

Agents update their beliefs according to the information carried by the past realizations of the growth rate X and estimate its unobservable drift $\phi(X,Y)$, (Section 1.3). In observable terms, they rewrite the dynamics of K as (cf. (1.14))

$$\frac{dK_t}{K_t} = \sigma_t \hat{\pi}_t^i dt + \sigma_t d\hat{W}_t^i, \tag{4.3}$$

where $\sigma_t = ||\xi(X)_t||$ is the volatility of X, which is observable, and $\hat{\pi}^i$ agent i's estimate of $\frac{\phi(X,Y)_t}{\sigma_t}$. Due to differences in prior beliefs, agents obtain different estimates $\hat{\pi}^i$ and different innovation processes $\hat{W}^i$. The latter are related as follows (cf. (1.16)):

$$\hat{W}_t^i = \hat{W}_t^j + \int_0^t (\hat{\pi}_u^j - \hat{\pi}_u^i)\, du\,. \tag{4.4}$$

The density processes

$$Z_t^i = \left.\frac{dP^i}{dP}\right|_{\mathcal{G}_t}$$

play an important role in the following. Therefore, it is useful to remember the stochastic differential equation

$$dZ_t^i = (\hat{\pi}_t^i - \hat{\pi}_t) Z_t^i d\hat{W}_t\,. \tag{4.5}$$

obtained in Lemma 1.3.4, (1.17).

4.2 Equilibrium

Under the reference measure P, the utility function of agent i has a von Neumann-Morgenstern representation with the state-dependent felicity function

$$v^i((\omega, t), d) = f(t) Z_t^i(\omega) \log(d)\,,$$

compare (1.8), page 14.

Theorem 4.2.1. *For $\lambda \gg 0$ the preferences of the representative agent are represented by the utility function*

$$\bar{U}(c, \lambda) = E^\lambda \int_0^T f(t) \log(c_t)\, dt$$

where the expectation is taken with respect to the measure

$$P^\lambda = \sum_i \lambda^i P^i .$$

The corresponding Pareto efficient allocation is

$$c^i(\lambda)_t = \frac{\lambda^i Z_t^i}{Z_t^\lambda} K_t , \tag{4.6}$$

with

$$Z_t^\lambda = \sum_j \lambda^j Z_t^j = \left. \frac{dP^\lambda}{dP} \right|_{\mathcal{G}_t} .$$

Proof. By Lemma 1.4.1, one has to solve

$$v(t,x,\alpha) = \max_{\sum y^i = x} \sum_i \alpha^i f(t) \log(y^i) . \tag{4.7}$$

The first order conditions are

$$\frac{\alpha^i f(t)}{y^i} = \nu$$

for some Lagrange multiplier ν. Using the constraint $\sum y^i = x$, this yields

$$y^i(t,x,\alpha) = \frac{\alpha^i}{\sum_j \alpha^j} x .$$

By (1.23),

$$c^i(\lambda)_t = y^i(t, K_t, (\lambda^i Z_t^i)_{i=1,\dots,I}) ,$$

hence (4.6) is proved.

Plugging in the optimal values of (y^i), the value function results as

$$\begin{aligned} v(t,x,\alpha) &= \sum_i \alpha^i f(t) \log\left(\frac{\alpha^i}{\sum_j \alpha^j} x\right) \\ &= \sum_i \alpha^i f(t) \log x + \sum_i \alpha^i f(t) \log\left(\frac{\alpha^i}{\sum_j \alpha^j}\right) . \end{aligned}$$

One version of the utility function of the representative agent is therefore

$$\tilde{U}(c,\lambda) = E\int_0^T \sum_i \lambda^i Z_t^i f(t)\log(c_t)\,dt + E\int_0^T \sum_i \lambda^i Z_t^i f(t)\log\left(\frac{\lambda^i Z_t^i}{\sum_j \lambda^j Z_t^j}\right)dt\,.$$

The second summand plays no role in representing the preference ordering and can be omitted. The preference ordering is therefore also represented by

$$\bar{U}(c,\lambda) = E\int_0^T \sum_i \lambda^i Z_t^i f(t)\log(c_t)\,dt\,.$$

Note that $Z_t^\lambda := \sum_i \lambda^i Z_t^i$ is the density of the probability measure $P^\lambda = \sum_i \lambda^i P^i$ with respect to the reference measure P. Hence,

$$\bar{U}(c,\lambda) = E^\lambda \int_0^T f(t)\log(c_t)\,dt\,.$$

There is a close analogy to the result obtained in Chapter 2 about the Preferred Habitat Theory. There, the consumption share of an agent in a Pareto optimum is determined by the likelihood fraction

$$\frac{\lambda^i f^i(t)}{f^\lambda(t)}$$

of time preference densities, see Theorem 2.3.1. Here, it is the corresponding likelihood fraction of beliefs,

$$\frac{\lambda^i Z^i(t)}{Z^\lambda(t)}\,,$$

which determines the consumption share of an agent.

Note that the share does not depend on the objective or reference measure P - only the "market belief" $P^\lambda = \sum_i \lambda^i P^i$ plays a role. The more favorable the observable data are for an agent's belief, compared to the beliefs of the others, the higher is her share.

The personal estimates $\hat{\pi}^i$ of $\pi_t = \frac{\phi(X,Y)_t}{\sigma_t}$ have already been introduced. The next lemma provides a link between these estimates and the "market estimate"

$$\hat{\pi}_t^\lambda := E^\lambda \left[\pi_t | \mathcal{G}_t \right].$$

Lemma 4.2.1. *Let $D \in \bigcap_i \mathcal{L}^1(P^i)$. Then $D \in \mathcal{L}^1(P^\lambda)$ and*

$$E^\lambda \left[D | \mathcal{G}_t \right] = \sum_i \frac{\lambda^i Z_t^i}{Z_t^\lambda} E^i \left[D | \mathcal{G}_t \right].$$

In particular, the market estimate $\hat{\pi}^\lambda$ is a convex combination of the personal estimates $\hat{\pi}^i$:

$$\hat{\pi}_t^\lambda = \sum_i \frac{\lambda^i Z_t^i}{Z_t^\lambda} \hat{\pi}_t^i.$$

Proof. Let $A \in \mathcal{G}_t$. The following chain of equalities holds for all integrable random variables D, especially for π_t:

$$\begin{aligned}
\int_A D dP^\lambda &= \sum_i \lambda^i \int_A D dP^i \\
&= \sum_i \lambda^i \int_A E^i \left[D | \mathcal{G}_t \right] dP^i \\
&= \sum_i \int_A \frac{\lambda^i Z_t^i}{Z_t^\lambda} E^i \left[D | \mathcal{G}_t \right] dP^\lambda .
\end{aligned}$$

The volatility coefficient of the density process Z^i is given by the deviation of the personal estimate $\hat{\pi}^i$ from the reference estimate $\hat{\pi}$. It comes as no surprise then, that the same holds true for the volatility coefficient of Z^λ:

Lemma 4.2.2. *Z^λ is a stochastic exponential and satisfies*

$$dZ_t^\lambda = \left(\hat{\pi}_t^\lambda - \hat{\pi}_t \right) Z_t^\lambda d\hat{W}_t,$$

or equivalently,

$$Z_t^\lambda = \exp\left(\int_0^t \left(\hat{\pi}_u^\lambda - \hat{\pi}_u \right) d\hat{W}_u - \frac{1}{2} \int_0^t \left(\hat{\pi}_u^\lambda - \hat{\pi}_u \right)^2 du \right).$$

Proof. With the use of the differential equation (4.5) for Z^i, one has

$$\begin{aligned} dZ_t^\lambda &= \sum_i \lambda^i dZ_t^i \\ &= \sum_i \lambda^i \left(\hat{\pi}_t^i - \hat{\pi}_t\right) Z_t^i d\hat{W}_t \\ &= \left(\sum_i \frac{\lambda^i Z_t^i}{Z_t^\lambda} (\hat{\pi}_t^i - \hat{\pi}_t)\right) Z_t^\lambda d\hat{W}_t \,, \end{aligned}$$

which by Lemma 4.2.1 is equal to

$$= \left(\hat{\pi}_t^\lambda - \hat{\pi}_t\right) Z_t^\lambda d\hat{W}_t \,.$$

Armed with the tools of the preceding lemmata, one can now proceed to derive the equilibrium of the economy.

Theorem 4.2.2. *An Arrow-Radner equilibrium with complete markets exists and is characterized by the vector $\lambda = s$ of initial shares the agents own of the stock.*

The equilibrium asset price is independent of the heterogeneity parameter s and given by

$$S_t = \frac{1 - F(t)}{f(t)} K_t \,. \tag{4.8}$$

The market price of risk is

$$p_t^s = \sigma_t - (\hat{\pi}_t^s - \hat{\pi}_t) \tag{4.9}$$

and the short rate is

$$r_t^s = \rho(t) + \sigma_t \hat{\pi}_t^s - \sigma_t^2 \,. \tag{4.10}$$

Proof. I show first, that the the functional $\Psi(s)$, given by (1.24), and the efficient allocation $c(s)$, which corresponds to the vector s, form an Arrow-Debreu equilibrium. For this, it suffices to show, that $c^i(s)$ is budget-feasible at prices $\Psi(s)$ for agent i,

$$\Psi(s)(c^i(s)) = \Psi(s)(e^i).$$

By (4.6), one has

$$\begin{aligned}
\Psi(s)(c^i(s)) &= E\int_0^T c^i(s)_u \psi(s)_u \, dt \\
&= E\int_0^T \frac{s^i Z_u^i}{\sum_j s^j Z_u^j} K_u f(u) Z_u^s K_u^{-1} \, dt \\
&= s^i .
\end{aligned}$$

On the other hand,

$$\begin{aligned}
\Psi(s)(e^i) &= s^i \Psi(s)(K) \\
&= s^i E\int_0^T K_u f(u) Z_u^s K_u^{-1} \, dt \\
&= s^i .
\end{aligned}$$

Hence, $(\Psi(s), (c^i(s))_i)$ is an Arrow-Debreu equilibrium.
By (1.29), the candidate for the equilibrium stock price is

$$\begin{aligned}
S_t &:= \frac{1}{\psi_t} E\left[\int_t^T K_u \psi_u \, du \,\middle|\, \mathcal{G}_t \right] \\
&= \frac{K_t}{f(t) Z_t^s} E\left[\int_t^T f(u) Z_u^s \, du \,\middle|\, \mathcal{G}_t \right] \\
&= \frac{1 - F(t)}{f(t)} K_t .
\end{aligned}$$

The market price of risk and the short rate are contained in the dynamics of the consumption price process $\psi_t = \psi(s)_t$:

$$\begin{aligned}
d\log \psi_t &= d\log f(t) + d\log Z_t^s - d\log K_t \\
&= -\rho(t) dt + (\hat{\pi}_t^s - \hat{\pi}_t) \, d\hat{W}_t - \frac{1}{2} (\hat{\pi}_t^s - \hat{\pi}_t)^2 \, dt \\
&\quad - \sigma_t d\hat{W}_t + \left(\frac{1}{2} \sigma_t^2 - \sigma_t \hat{\pi}_t \right) dt ,
\end{aligned}$$

where one uses the definition of the rate of time preference, Lemma 4.2.2 and (4.3).

On the other hand, the short rate and the market price of risk satisfy (compare (1.28))

$$d\log\psi_t = \left(\frac{1}{2}(p^s)_t^2 - r_t^s\right)dt - p_t^s d\hat{W}_t . \qquad (4.11)$$

Comparing terms in (4.11) and (4.11) yields

$$p_t^s = \sigma_t - (\hat{\pi}_t^s - \hat{\pi}_t)$$

and

$$\begin{aligned} r_t^s &= \rho(t) + \sigma_t\hat{\pi}_t - \frac{1}{2}\sigma_t^2 + \frac{1}{2}\left(\hat{\pi}_t^s - \hat{\pi}_t\right)^2 - \frac{1}{2}\left(\sigma_t - (\hat{\pi}_t^s - \hat{\pi}_t)\right)^2 \\ &= \rho(t) + \sigma_t\hat{\pi}_t^s - \sigma_t^2 . \end{aligned}$$

It suffices to check the condition (1.31) to show that the asset market given by the stock and the money market account is complete. Then the existence of an Arrow-Radner equilibrium follows from Theorem 1.5.1 and Theorem 1.5.2. The deflated gain process is

$$\begin{aligned} G_t &= \int_0^t K_u\psi_u du + S_t\psi_t \\ &= \int_0^t f(u)Z_u^s du + (1 - F(t))Z_t^s . \end{aligned}$$

Its dynamics are

$$\begin{aligned} dG_t &= f(t)Z_t^s dt - Z_t^s dF(t) + (1 - F(t))dZ_t^s \\ &= (1 - F(t))dZ_t^s \\ &= (1 - F(t))(\hat{\pi}_t^s - \hat{\pi}_t)Z_t^s d\hat{W}_t . \end{aligned} \qquad (4.12)$$

With the ingredients at hand, one checks now the condition (1.31), namely

$$p^s S\psi \neq \gamma G \quad P \otimes dt - a.e.,$$

which is equivalent to

$$(\sigma_t - (\hat{\pi}_t^s - \hat{\pi}_t))\,(1 - F(t))Z_t^s \neq$$
$$-(1 - F(t)(\hat{\pi}_t^s - \hat{\pi}_t)Z_t^s \quad P \otimes dt - a.e.,$$

or

$$\sigma_t Z_t^s \neq 0 \quad P \otimes dt - a.e.$$

But σ_t and Z_t^s are strictly positive a.e., since the beliefs are mutually equivalent and the volatility does never vanish by Assumption (1.4).

It is remarkable that the stock price does not depend on agents' belief. This is due to logarithmic felicity, which entails that the density Z^s, which reflects the beliefs of the agents, drops out in the calculations.

Beliefs do have an impact on the short rate through the estimate $\sigma_t \hat{\pi}^s$ of the unobservable drift $\phi(X, Y)_t$. Again I stress that it is the estimate under the market belief P^s which sets the short rate, not the estimate under the reference measure P. In the homogeneous world populated by type i-agents, the market estimate is equal to the personal estimate $\sigma_t \hat{\pi}_t^i$ of agent i. The following analogue to the conjecture of Dumas, Corollary 2.3.1 in Chapter 2, is a corollary to Lemma 4.2.1:

Theorem 4.2.3. *The short rate r^s in the mixed economy is a convex combination of the short rates r^i which prevail in homogeneous economies,*

$$r_t^s = \sum_i \frac{s^i Z_t^i}{Z_t^s} r_t^i. \tag{4.13}$$

The same holds true for the market price of risk:

$$p_t^s = \sum_i \frac{s^i Z_t^i}{Z_t^s} p_t^i.$$

Proof. Directly from (4.9), (4.10) and Lemma 4.2.1.

Bond prices $B^{s,T'}$ in the mixed economy are also weighted averages of the bond prices $B^{i,T'}$ of homogeneous economies.

Theorem 4.2.4. *Equilibrium bond prices are convex combinations of bond prices prevailing in homogeneous economies:*

$$B_t^{s,T'} = E^s\left[\frac{f(T')K_t}{f(t)K_{T'}}\,\middle|\, \mathcal{G}_t\right] = \sum_i \frac{s^i Z_t^i}{Z_t^s} B_t^{i,T'} .$$

The forward rates $f_t^{s,T'}$ are also convex combinations of the forward rates $f_t^{i,T'}$ which prevail in the homogeneous economies:

$$f_t^{s,T'} = \sum_i w_t^{i,T'} f_t^{i,T'} ,$$

where the weights $w_t^{i,T'}$ are given by

$$w_t^{i,T'} = \frac{s^i Z_t^i B_t^{i,T'}}{\sum_j s^j Z_t^j B_t^{j,T'}} .$$

Proof. By the general pricing formula (1.44),

$$B_t^{s,T'} = E\left[\frac{\psi_{T'}^s}{\psi_t^s}\,\middle|\, \mathcal{G}_t\right] = E\left[\frac{f(T')K_t Z_{T'}^s}{f(t)K_{T'} Z_t^s}\,\middle|\, \mathcal{G}_t\right] .$$

Since Z^s is the density of P^s with respect to P, it follows by Bayes' rule

$$B_t^{s,T'} = E^s\left[\frac{f(T')K_t}{f(t)K_{T'}}\,\middle|\, \mathcal{G}_t\right] .$$

By Lemma 4.2.1,

$$E^s\left[\frac{f(T')K_t}{f(t)K_{T'}}\,\middle|\, \mathcal{G}_t\right] = \sum_i \frac{s^i Z_t^i}{Z_t^\lambda} E^i\left[\frac{f(T')K_t}{f(t)K_{T'}}\,\middle|\, \mathcal{G}_t\right] ,$$

hence

$$B_t^{s,T'} = \sum_i \frac{s^i Z_t^i}{Z_t^s} B_t^{i,T'} . \tag{4.14}$$

The forward rates are by definition

$$f_t^{s,T'} = -\frac{\frac{\partial}{\partial T'} B_t^{s,T'}}{B_t^{s,T'}} .$$

Using (4.14), one has

$$\begin{aligned} f_t^{s,T'} &= -\frac{\sum_i s^i Z_t^i B_t^{i,T'} \frac{\frac{\partial}{\partial T'} B_t^{i,T'}}{B_t^{i,T'}}}{Z_t^s B_t^{s,T'}} \\ &= \sum_i w_t^{i,T'} f_t^{i,T'} . \end{aligned}$$

The price of a bond with maturity T' is given by the expected value of the fraction of marginal felicities $\frac{f(T)K_t}{f(t)K_{T'}}$ as in the standard CCAPM. The difference is that the expectation is taken with respect to the market's belief P^s, not with respect to the reference measure P.

The preceding Theorems 4.2.3 and 4.2.4 clarify completely how the differences in beliefs among agents influence the term structure. The short rate, the starting point of the yield curve, is in between the short rates of the homogeneous worlds. Moreover, the whole yield curve is bounded by the yield curves of the homogeneous worlds. The highest (lowest) yield of the homogeneous worlds is a upper (lower) bound for the yield in the mixed world.

Agents have different demands since they differ in beliefs. They have, therefore, an incentive to trade with another and one is curious to know, who behaves how in equilibrium? What kind of asset does the bull prefer, what kind is appropriate for the bear? Here is the answer.

Theorem 4.2.5. *Agent i holds*

$$\theta_t^{1i} = \frac{s^i Z_t^i}{Z_t^s} \left(1 + \frac{\hat{\pi}_t^i - \hat{\pi}_t^s}{\sigma_t} \right) \tag{4.15}$$

shares of the stock and

$$\theta_t^{2i} = -\frac{s^i Z_t^i(1-F(t))\left(\hat{\pi}_t^i - \hat{\pi}_t^s\right)}{\sigma_t H_t} \tag{4.16}$$

shares of the money account.

Proof. The argument is as outlined in the proof of Theorem 1.5.1.

The deflated value of agent i's portfolio is

$$\begin{aligned} V_t^i &= E\left[\int_t^T c_u^i \psi_u du \,\middle|\, \mathcal{G}_t\right] \\ &= s^i Z_t^i(1-F(t)) . \end{aligned} \tag{4.17}$$

Its dynamics are, using (4.5),

$$dV_t^i = -s^i Z_t^i f(t)dt + s^i Z^i(\hat{\pi}_t^i - \hat{\pi}_t)d\hat{W}_t .$$

The dynamics of the gain process G are, cf. (4.12),

$$dG_t = (1-F(t))(\hat{\pi}_t^s - \hat{\pi}_t)Z_t^s d\hat{W}_t . \tag{4.18}$$

One has therefore to solve the linear system

$$\begin{aligned} \theta_t^{1i}(1-F(t))Z_t^s \quad &+ \theta_t^{2i}H_t = \quad s^i Z_t^i(1-F(t)) \\ \theta_t^{1i}(1-F(t))(\hat{\pi}_t^s - \hat{\pi}_t)Z_t^s &- \theta_t^{2i}p_t^s H_t = s^i Z_t^i(1-F(t))(\hat{\pi}_t^i - \hat{\pi}_t). \end{aligned}$$

Its determinant is

$$\begin{aligned} det &= -(1-F(t))Z_t^s p_t^s H_t - H_t(1-F(t))(\hat{\pi}_t^s - \hat{\pi}_t)Z_t^s \\ &= -(1-F(t))H_t Z_t^s\left(p_t^s + \hat{\pi}_t^s - \hat{\pi}_t\right) , \end{aligned}$$

which by (4.9) is equal to

$$= -(1-F(t))H_t Z_t^s \sigma_t .$$

The solution for θ^{1i} is

$$\begin{aligned} \theta_t^{1i} &= \frac{-s^i Z_t^i(1-F(t))p_t^s H_t - H_t s^i Z_t^i(1-F(t))(\hat{\pi}_t^i - \hat{\pi}_t)}{det} \\ &= \frac{s^i Z_t^i\left(p_t^s + \hat{\pi}_t^i - \hat{\pi}_t\right)}{\sigma_t Z_t^s} \end{aligned}$$

which is equal to (4.15) . The same calculation yields θ_t^{2i}.

The fraction of wealth an agent invests in the stock is

$$\frac{\theta_t^{1i} S_t \psi_t}{V_t^i} = 1 + \frac{\hat{\pi}_t^i - \hat{\pi}_t^s}{\sigma_t}.$$

An agent invests more than 100% in the stock and borrows in the money market if and only if her estimate of the drift is higher than the market's estimate, $\hat{\pi}^i > \hat{\pi}^s$. Bulls, so to speak, invest in the risky asset, whereas bears invest in the money market account and sell shares of the asset.

In the proof of the last theorem, the wealth of an agent

$$V_t^i = s^i Z_t^i (1 - F(t))$$

has been calculated (cf. (4.17)). Aggregate wealth is therefore

$$V_t = \sum_i V_t^i = Z_t^s (1 - F(t)).$$

Detemple and Murthy [15] derive the result that short rate and bond prices in the mixed economy are weighted averages of the corresponding quantities in homogeneous economies where the weights are the fractions of relative wealth

$$\frac{s^i V_t^i}{V_t}.$$

The same holds here because relative wealth is equal to the likelihood fraction,

$$\frac{s^i V_t^i}{V_t} = \frac{s^i Z_t^i}{Z_t^s}.$$

4.3 Two Examples

4.3.1 Unobservable Constant Drift

The first example, in which the yield curve is explicitly computable, is the case of an unobservable constant drift of the growth rate,

$$dX_t = Y_t dt + dW_t$$
$$Y_t = \eta \quad \text{all } t$$

as in Chapter 3. There are two agents with normal prior beliefs

$$\mathcal{L}\left(\eta | P^i\right) = N(m^i, 1)$$

and $m^1 > m^2$. Thus, the first agent is more optimistic than the second about the mean growth of the economy's output. The estimate of η is

$$\hat{\pi}_t^i = E^i\left[\eta | \mathcal{G}_t\right] = \frac{m^i + X_t}{1 + t},$$

compare (3.4). The short rate in homogeneous economies is, setting the rate of time preference $\rho = 0$ for simplicity,

$$\hat{r}_t^i = \hat{\pi}_t^i - 1.$$

The yield curves in the homogeneous worlds are linearly decreasing,

$$\widehat{y_t^{i,t+\tau}} = \widehat{r_t^i} - \frac{\tau}{2(1+t)},$$

compare (3.16). The yield curve in the mixed economy is in between the curves of the homogeneous economies and is given by

$$\begin{aligned}\widehat{y_t^{s,t+\tau}} = -\frac{1}{\tau} \log \Bigg(&\alpha \exp\left(-\tau\left[\widehat{r_t^1} - \frac{\tau}{2(1+t)}\right]\right) \\ &+ (1-\alpha) \exp\left(-\tau\left[\widehat{r_t^2} - \frac{\tau}{2(1+t)}\right]\right)\Bigg)\end{aligned}$$

where the weight $\alpha = \frac{s^i Z_t^i}{Z_t^s}$ is the likelihood fraction of beliefs which is independent of time to maturity τ. The formula simplifies to

$$\widehat{y_t^{s,t+\tau}} = \widehat{r_t^2} - \frac{\tau}{2(1+t)} - \frac{1}{\tau} \log\left(1 + \alpha\left[e^{-\tau\frac{m^1 - m^2}{1+t}} - 1\right]\right).$$

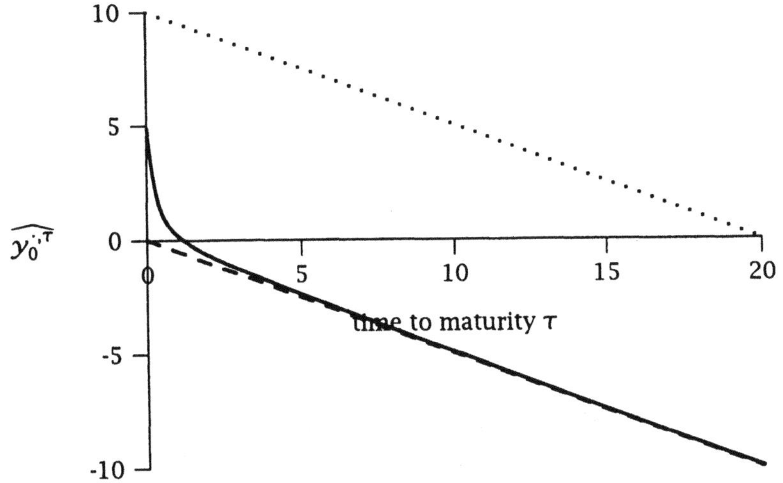

Fig. 4.1. The yield curves when the growth rate is constant. The parameter values are $m^1 = 11, m^2 = 1$ and $\alpha = \frac{1}{2}$. The yield curves of the homogeneous economies are represented in dashes and dots, resp. The yield curve of the mixed economy is the solid line.

The second summand is positive and tends to 0 as maturity τ reaches infinity. Asymptotically, the yield curve in the mixed world approaches therefore the lower curve $\widehat{y_t^{2,t+\tau}}$ of the more pessimistic agent. A picture is given in Figure 4.1.

Heterogeneity of agents does not remove the astonishing result of Chapter 3 because the yield curve is also decreasing in the mixed world.

4.3.2 Stationary Unobservable Drift

The second example has two independent Brownian motions W^1 and W^2. It illustrates, therefore, also the general theme of Chapter 1, namely how imperfect information reduces the observable noise to a one-dimensional Brownian motion $\hat{W}$, which allows to construct a complete asset market. Similar dynamics are studied in Feldman [24] in the framework of a homogeneous production economy.

Again, the growth rate X has an unobservable drift

$$dX_t = Y_t dt + dW_t^1 ,$$

but Y is, in contrast to the preceding example, a stationary process of the form

$$dY_t = -\mu Y_t dt + dW_t^2, \quad Y_0 = \eta.$$

Then the estimate $\hat{\pi}_t^i = E^i[Y_t | \mathcal{G}_t]$ solves the stochastic differential equation

$$d\hat{\pi}_t^i = -\mu \hat{\pi}_t^i dt + \gamma(t) d\hat{W}_t^i,$$

where $\hat{W}_t^i = X_t - \int_0^t \hat{\pi}_u^i du$ is the innovation process of agent i. $\gamma(t)$ is the conditional variance of Y_t given the observed past $\mathcal{G}_t$. By a result of filtering theory, γ is a deterministic function and solves the differential equation

$$\gamma'(t) = 1 - 2\mu\gamma(t) - \gamma^2(t),$$

compare Liptser, Shiryaev [39], Theorem 10.1.

The short rate is, as in the preceding example,

$$\hat{r}_t^i = \hat{\pi}_t^i - 1.$$

Remember that the rate of time preference is set to zero for simplicity.

Feldman [24] solves for the bond prices in the homogeneous world with the use of partial differential equations. I offer a more direct way using the fact that the integral over an autoregressive process is related to the process itself.

Lemma 4.3.1. *Bond prices in homogeneous economies are given by*

$$B_t^{i,t+\tau} = \exp\left(\frac{1}{2}\tau - \frac{1 - e^{-\mu\tau}}{\mu}\hat{\pi}_t^i + \frac{1}{2}\int_t^{t+\tau}\left\{\frac{1 - e^{-\mu(t+\tau-s)}}{\mu}\gamma(s) + 1\right\}^2 ds\right).$$

Proof. Bond prices are given by the fraction of marginal felicities

$$
\begin{aligned}
B_t^{t+\tau} &= E\left[\frac{K_t}{K_{t+\tau}}\,\middle|\, \mathcal{G}_t\right] \\
&= E\left[\exp\left(-(X_{t+\tau}-X_t)+\frac{1}{2}\tau\right)\,\middle|\, \mathcal{G}_t\right] \\
&= e^{\frac{1}{2}\tau} E\left[\exp\left(-\int_t^{t+\tau}\hat{\pi}_u du - (\hat{W}_{t+\tau}-\hat{W}_t)\right)\,\middle|\, \mathcal{G}_t\right],
\end{aligned}
$$

dropping the superscript i. Due to the autoregressive structure of $\hat{\pi}$, the integral over the the process is related to the process itself:

$$
-\int_t^{t+\tau}\hat{\pi}_u du = \frac{\hat{\pi}_{t+\tau}-\hat{\pi}_t}{\mu} - \int_t^{t+\tau}\frac{\gamma(s)}{\mu}d\hat{W}_u .
$$

Moreover, $\hat{\pi}$ is explicitly given by

$$
\hat{\pi}_{t+\tau} = e^{-\mu\tau}\left(\hat{\pi}_t + \int_t^{t+\tau} e^{\mu(s-t)}\gamma(s)d\hat{W}_s\right).
$$

This leads to

$$
\begin{aligned}
-\int_t^{t+\tau}\hat{\pi}_u du &= -\frac{1-e^{-\mu\tau}}{\mu}\hat{\pi}_t + \int_t^{t+\tau}\frac{e^{-\mu(t+\tau-s)}\gamma(s)}{\mu}d\hat{W}_s \\
&\quad -\int_t^{t+\tau}\frac{\gamma(s)}{\mu}d\hat{W}_s .
\end{aligned}
$$

Bond prices can therefore be written as a conditional Laplace transform of a Brownian integral with deterministic integrand:

$$
\begin{aligned}
B_t^{t+\tau} &= e^{\frac{1}{2}\tau - \frac{1-e^{-\mu\tau}}{\mu}\hat{\pi}_t} E\left[e^{-\int_t^{t+\tau}\left(\frac{1-e^{-\mu(t+\tau-s)}}{\mu}\gamma(s)+1\right)d\hat{W}_s}\,\middle|\, \mathcal{G}_t\right] \\
&= e^{\frac{1}{2}\tau - \frac{1-e^{-\mu\tau}}{\mu}\hat{\pi}_t + \frac{1}{2}\int_t^{t+\tau}\left(\frac{1-e^{-\mu(t+\tau-s)}}{\mu}\gamma(s)+1\right)^2 ds} .
\end{aligned}
$$

References

1. A.B. Abel. Asset prices under habit formation and catching up with the Joneses. *American Economic Review*, 80:38–42, 1990.
2. A. Araujo and P. K. Monteiro. Generic non-existence of equilibria in finance models. *Journal of Mathematical Economics*, 20:489–499, 1991.
3. K.J. Arrow. Le rôle des valeurs boursières pour la repartition la meilleure des risques. In *Econométrie*, Paris, 1953. Centre National de la Recherche Scientifique.
4. K.J. Arrow. *Essays in the Theory of Risk Bearing*. Markham Publishing Co., Chicago, 1971.
5. K.J. Arrow and G. Debreu. Existence of equilibria for a competitive economy. *Econometrica*, 22:265–290, 1954.
6. R. Aumann. Agreeing to disagree. *The Annals of Statistics*, 4:1236–1239, 1976.
7. G.S. Bakshi and Z. Chen. An alternative valuation model for contingent claims. *Journal of Financial Economics*, 44(1):123–165, 1997.
8. F. Black and M. Scholes. The pricing of options and corporate liabilities. *Journal of Political Economy*, 81:637–654, 1973.
9. G.M. Constantinides. Habit formation: A resolution of the equity premium puzzle. *Journal of Political Economy*, 98:519 543, 1990.
10. J.C. Cox, J.E. Ingersoll, and S.R. Ross. A reexamination of traditional hypotheses about the term structure of interest rates. *Journal of Finance*, 36:769–799, 1981.
11. J.C. Cox, J.E. Ingersoll, and S.R. Ross. A theory of the term structure of interest rates. *Econometrica*, 53:385–408, 1985.
12. R. Dana. Existence and uniqueness of equilibria when preferences are additively separable. *Econometrica*, 61:953–957, 1993.
13. R. Dana and M. Pontier. On existence of an Arrow-Radner equilibrium in the case of complete markets. A remark. *Mathematics of Operations Research*, 17:148–163, 1992.
14. J. Detemple. Asset pricing in a production economy with incomplete information. *Journal of Finance*, 41:383–391, 1986.
15. J. Detemple and S. Murthy. Intertemporal asset pricing with heterogeneous beliefs. *Journal of Economic Theory*, 62:294–320, 1994.
16. J. Detemple and F. Zapatero. Asset prices in an exchange economy with habit formation. *Econometrica*, 59:1633–1657, 1991.
17. L. U. Dothan. On the term structure of interest rates. *Journal of Financial Economics*, 6:59–69, 1978.
18. M. U. Dothan and D. Feldman. Equilibrium interest rates and multiperiod bonds in a production-exchange economy. *Journal of Finance*, 41:369–382, 1986.

19. D. Duffie. *Dynamic Asset Pricing Theory.* Princeton University Press, Princeton, 1992.
20. D. Duffie and L.G. Epstein. Stochastic differential utility. *Econometrica,* 60:353–394, 1992.
21. D. Duffie and C.F. Huang. Implementing Arrow-Debreu equilibria by continuous trading of few long-lived securities. *Econometrica,* 53:1337–1356, 1985.
22. D. Duffie, M. Schroder, and C. Skiadas. A term structure model with preferences for the timing of resolution of uncertainty. *Economic Theory,* 9(1):3–22, 1997.
23. B. Dumas. Two-person dynamic equilibrium in the capital market. *Review of Financial Studies,* 2:157–188, 1989.
24. D. Feldman. The term structure of interest rates in a partially observable economy. *Journal of Finance,* 44:789–812, 1989.
25. J.M. Harrison and D. Kreps. Martingales and arbitrage in multiperiod security markets. *Journal of Economic Theory,* 20:381–408, 1979.
26. J.M. Harrison and S.R. Pliska. Martingales and stochastic integrals in the theory of continuous trading. *Stochastic Processes and Their Applications,* 11:215–260, 1981.
27. J. Hicks. *Value and Capital.* Oxford University Press, 1939.
28. A. Hindy, C. Huang, and D. Kreps. On intertemporal preferences in continuous time: The case of certainty. *Journal of Mathematical Economics,* 21(5):401–440, 1992.
29. J. Hull and A. White. Pricing interest rate derivative securities. *Review of Financial Studies,* 3:573–592, 1990.
30. G. Kallianpur. *Stochastic Filtering Theory.* Springer-Verlag, New York, 1980.
31. I. Karatzas. *Lectures on the Mathematics of Finance.* CRM Monograph Series. American Mathematical Society, Providence, R.I., 1997.
32. I. Karatzas, J. P. Lehoczky, and S. E. Shreve. Optimal portfolio and consumption decisions for a "small investor" on a finite horizon. *SIAM Journal of Control and Optimization,* 25:1557–1586, 1987.
33. I. Karatzas, J. P. Lehoczky, and S. E. Shreve. Existence and uniqueness of multi-agent equilibrium in a stochastic, dynamic consumption/investment model. *Mathematics of Operations Research,* 15:80–128, 1990.
34. I. Karatzas and S. E. Shreve. *Brownian Motion and Stochastic Calculus.* Springer-Verlag, New York, 1991.
35. J.M. Keynes. *The General Theory of Employment, Interest, and Money.* Macmillan Co. Ltd., London, 1936.
36. M.S. Kimball. Precautionary saving in the small and in the large. *Econometrica,* 58:53–73, 1990.
37. D. Kreps. A note on 'fulfilled expectations' equilibria. *Journal of Economic Theory,* 14:32–43, 1977.
38. D. Kreps and E. Porteus. Temporal resolution of uncertainty and dynamic choice theory. *Econometrica,* 46:185–200, 1978.
39. R. S. Liptser and A. N. Shiryayev. *Statistics of Random Processes I and II.* Springer-Verlag, New York, 1977.
40. R.E. Lucas. Asset prices in an exchange economy. *Econometrica,* 46:1429–1445, 1978.
41. F.A. Lutz. Term structure of interest rates. *Quarterly Journal of Economics,* pages 36–63, 1940.
42. B.G. Malkiel. Expectation bond prices and the term structure of interest rates. *Quarterly Journal of Economics,* pages 197–218, 1962.

43. A. Mas-Colell, M. D. Whinston, and J. R. Green. *Microeconomic Theory*. Oxford University Press, 1995.
44. A. Mas-Colell and W. Zame. Equilibrium theory in infinite dimensional spaces. In W. Hildenbrand and H. Sonnenschein, editors, *Handbook of Mathematical Economics*, volume 4, pages 1835-1898. Elsevier Science Publishers B.V., North-Holland, Amsterdam, 1991.
45. R. Mehra and E. Prescott. The equity premium: A puzzle. *Journal of Monetary Economics*, 15:145-161, 1985.
46. R. C. Merton. A dynamic general equilibrium model of the asset market and its application to the pricing of the capital structure of the firm. Unpublished Working Paper, Sloan School of Management, MIT, 1970.
47. R. C. Merton. Optimum consumption and portfolio rules in a continuous-time model. *Journal of Economic Theory*, 3:373-413, 1971.
48. R. C. Merton. *Continuous-Time Finance*. Blackwell Publishers, 1990.
49. P. Milgrom and N. Stokey. Information, trade and common knowledge. *Journal of Economic Theory*, 26:17-27, 1982.
50. F. Modigliani and R. Sutch. Innovations in interest rate policy. *American Economic Review*, pages 178-197, May 1966.
51. T. Negishi. Welfare economics and the existence of an equilibrium for a competitive economy. *Metroeconomica*, 12:92-97, 1960.
52. J.W. Pratt. Risk aversion in the small and in the large. *Econometrica*, 32:122-136, 1964.
53. R. Radner. Existence of equilibrium of plans, prices and price expectations in a sequence of markets. *Econometrica*, 40:289-303, 1972.
54. S. F. Richard. An arbitrage model of the term structure of interest rates. *Journal of Financial Economics*, 6:33-57, 1978.
55. L.C.G. Rogers. Which model for the term structure of interest rates should one use? In M. Davis, D. Duffie, W. Fleming, and S. Shreve, editors, *Mathematical Finance*, pages 92-115. Springer-Verlag, New York, 1995.
56. R.J. Shiller. The term structure of interest rates. In B.M. Friedman and F.H. Hahn, editors, *Handbook of Monetary Economics*, volume I, pages 627-722. Elsevier Science Publishers B.V., North-Holland, Amsterdam, 1990.
57. R. H. Strotz. Myopia and inconsistency in dynamic utility maximization. *Review of Economic Studies*, 23:165-180, 1956.
58. T.S. Sun. Real and nominal interest rates: A discrete-time model and its continuous-time limit. *Review of Financial Studies*, 5:582-612, 1992.
59. J. Tirole. On the possibility of speculation under rational expectations. *Econometrica*, 50(5):1163-1182, 1982.
60. H. R. Varian. *Microeconomic Analysis*. W.W. Norton and Co., 1992.
61. O. Vasiček. An equilibrium characterization of the term structure. *Journal of Financial Economics*, 5:177-188, 1977.
62. J. Wang. The term structure of interest rates in a pure exchange economy with heterogeneous investors. *Journal of Financial Economics*, 41:75-110, 1996.
63. T. Yamada and S. Watanabe. On the uniqueness of solutions of stochastic differential equations. *Journal Math. Kyoto University*, 11:155-167, 1971.

List of Figures

Contributions to Economics

Sabine Spangenberg
The Institutionalised Transformation of the East German Economy
1998. ISBN 3-7908-1103-3

Hagen Bobzin
Indivisibilities
1998. ISBN 3-7908-1123-8

Helmut Wagner (Ed.)
Current Issues in Monetary Economics
1998. ISBN 3-7908-1127-0

Peter Michaelis/Frank Stähler (Eds.)
Recent Policy Issues in Environmental and Resource Economics
1998. ISBN 3-7908-1137-8

Jessica de Wolff
The Political Economy of Fiscal Decisions
1998. ISBN 3-7908-1130-0

Georg Bol/
Gholamreza Nakhaeizadeh/
Karl-Heinz Vollmer (Eds.)
Risk Measurements, Econometrics and Neural Networks
1998. ISBN 3-7908-1152-1

Joachim Winter
Investment and Exit Decisions at the Plant Level
1998. ISBN 3-7908-1154-8

Bernd Meyer
Intertemporal Asset Pricing
1999. ISBN 3-7908-1159-9

Uwe Walz
Dynamics of Regional Integration
1999. ISBN 3-7908-1185-8

Michael Carlberg
European Monetary Union
1999. ISBN 3-7908-1191-2

Giovanni Galizzi/
Luciano Venturini (Eds.)
Vertical Relationships and Coordination in the Food System
1999. ISBN 3-7908-1192-0

Gustav A. Horn/
Wolfgang Scheremet/
Rudolf Zwiener
Wages and the Euro
1999. ISBN 3-7908-1199-8

Dirk Willer
The Development of Equity Capital Markets in Transition Economies
1999. ISBN 3-7908-1198-X

Karl Matthias Weber
Innovation Diffusion and Political Control of Energy Technologies
1999. ISBN 3-7908-1205-6

Heike Link et al.
The Costs of Road Infrastructure and Congestion in Europe
1999. ISBN 3-7908-1201-3

Simon Duindam
Military Conscription
1999. ISBN 3-7908-1203-X

Bruno Jeitziner
Political Economy of the Swiss National Bank
1999. ISBN 3-7908-1209-9

Irene Ring et al. (Eds.)
Regional Sustainability
1999. ISBN 3-7908-1233-1

Katharina Müller/Andreas Ryll/
Hans-Jürgen Wagener (Eds.)
Transformation of Social Security: Pensions in Central-Eastern Europe
1999. ISBN 3-7908-1210-2

Stefan Traub
Framing Effects in Taxation
1999. ISBN 3-7908-1240-4

Pablo Coto-Millán
Utility and Production
1999. 3-7908-1153-X

GPSR Compliance

The European Union's (EU) General Product Safety Regulation (GPSR) is a set of rules that requires consumer products to be safe and our obligations to ensure this.

If you have any concerns about our products, you can contact us on ProductSafety@springernature.com

In case Publisher is established outside the EU, the EU authorized representative is:

Springer Nature Customer Service Center GmbH
Europaplatz 3
69115 Heidelberg, Germany

Batch number: 10725973

Printed by Printforce, the Netherlands